Navigation: Science and Technology

Volume 18

This series *Navigation: Science and Technology (NST)* presents new developments and advances in various aspects of navigation - from land navigation, marine navigation, aeronautic navigation to space navigation; and from basic theories, mechanisms, to modern techniques. It publishes monographs, edited volumes, lecture notes and professional books on topics relevant to navigation - quickly, up to date and with a high quality. A special focus of the series is the technologies of the Global Navigation Satellite Systems (GNSSs), as well as the latest progress made in the existing systems (GPS, BDS, Galileo, GLONASS, etc.). To help readers keep abreast of the latest advances in the field, the key topics in NST include but are not limited to:

- Satellite Navigation Signal Systems
- GNSS Navigation Applications
- Position Determination
- Navigational instrument
- Atomic Clock Technique and Time-Frequency System
- X-ray pulsar-based navigation and timing
- Test and Evaluation
- User Terminal Technology
- Navigation in Space
- New theories and technologies of navigation
- Policies and Standards

This book series is indexed in **SCOPUS** and **EI Compendex** databases.

Zichen Huang

Acoustic and Signal-Based Local Positioning

Techniques for GPS-Denied Spaces

Springer

Zichen Huang
Zhejiang University
Hangzhou, Zhejiang, China

ISSN 2522-0454 ISSN 2522-0462 (electronic)
Navigation: Science and Technology
ISBN 978-981-95-6082-0 ISBN 978-981-95-6083-7 (eBook)
https://doi.org/10.1007/978-981-95-6083-7

This Springer imprint is published by the registered company Springer Nature Singapore Pte Ltd.
The registered company address is: 152 Beach Road, #21-01/04 Gateway East, Singapore 189721, Singapore

Preface

Indoor robotic systems are becoming increasingly prevalent across industry and research, from autonomous drones and mobile robots in warehouses to smart farming equipment in greenhouses. A fundamental enabler for these systems is precise indoor positioning, yet conventional satellite-based navigation (GPS/GNSS) is often ineffective once we move under a roof. This limitation has prompted the exploration of alternative positioning methods tailored for GPS-denied environments. Among the various approaches, including radio-frequency and vision-based systems, acoustic positioning has emerged as a compelling solution for indoor localization. Acoustic Positioning Systems leverage sound signals (typically ultrasonic) to measure distances and compute positions. They offer notable advantages such as the ability to achieve centimeter-level accuracy, the use of low-cost and readily available hardware components, and operation in a separate spectrum that avoids interference with radio networks. These characteristics make sound-based localization especially attractive for environments where other systems face challenges, providing a practical path to "indoor GPS" capabilities.

The motivation for this book stems from the authors' years of research and development in acoustic indoor positioning. In particular, this work has focused on Sound-based Spread Spectrum Local Positioning Systems (SSSLPS), a class of acoustic systems that use coded ultrasound signals to enhance robustness and precision. Through a series of projects applying SSSLPS in demanding settings like agricultural greenhouses, we have witnessed firsthand both the potential and the challenges of sound-based localization. Early prototypes demonstrated that, with proper design, an acoustic system could reliably attain positioning errors on the order of only a few centimeters over tens of meters, even in cluttered indoor environments. Such results confirmed that acoustic methods can meet the stringent accuracy requirements of applications like greenhouse robotics, where narrow aisles and delicate crops demand extremely precise navigation. These experiences not only validated the feasibility of acoustic positioning for indoor robots, but also revealed practical insights—from signal processing techniques to deployment considerations—that are not widely documented in existing literature. This book is our attempt to consolidate those insights and share the design principles, application examples, and future trends of acoustic positioning technology with the broader community.

Our intent is to provide a comprehensive and authoritative resource on indoor acoustic positioning without retracing the same ground covered by standard introductions to localization. While the book's Introduction lays out the technical background and comparative overview of various indoor positioning technologies, this Preface offers a brief contextualization of why a dedicated volume on acoustic methods is timely and valuable. In recent years, indoor localization has grown into a vibrant field, but acoustic techniques, despite their advantages, have remained a niche area. We believe that the maturation of SSSLPS research, along with increasing demand from industries like autonomous logistics and precision agriculture, makes this an opportune moment to present a focused treatment of acoustic positioning systems. By covering topics from fundamental physics and signal design to system integration and field applications, we aim to bridge the gap between theoretical underpinnings and practical engineering. The chapters that follow are organized to guide readers from foundational concepts to hands-on implementation and emerging developments, reflecting our conviction that effective learning occurs when theory is reinforced by real-world context.

This book is written for engineers, students, and industry professionals working in robotics, automation, and related fields who seek a deep understanding of indoor acoustic positioning. We have tailored the content to be accessible to graduate students and newcomers by introducing essential concepts of acoustics, signal processing, and localization in the early chapters, while also providing sufficient technical depth and references to serve as a reference manual for experienced researchers and practitioners. Our hope is that different types of readers will find value in these pages: a development engineer looking to implement a low-cost indoor navigation system, a graduate student exploring alternatives to radio-based tracking, or a technical manager evaluating solutions for a GPS-denied facility. To this end, we include not only explanations of algorithms and design choices, but also practical insights gained from field experiments, plus discussions on the applications and future trends (such as multi-robot systems and drone navigation) where acoustic positioning shows promise. We intend for the material to stand on its own, so that readers can grasp the key ideas even without extensive prior knowledge, yet we cite key literature throughout for those interested in digging deeper.

Finally, we would like to express our sincere gratitude to Prof. Naoshi Kondo for his invaluable guidance over the past few years regarding our research on SSSLPS. Prof. Kondo's mentorship and expertise in sound-based positioning have greatly influenced our work and shaped the direction of this book. From conceiving robust spread-spectrum signaling strategies to refining system prototypes in greenhouse trials, his insights have been instrumental at every step. We are deeply thankful for his encouragement, technical advice, and the many discussions that helped sharpen our understanding. It is our privilege to acknowledge his contribution and support here. We also extend our appreciation to colleagues and collaborators in our research group who assisted in various experiments and provided feedback during the preparation of the manuscript.

In closing, we hope that Acoustic Positioning Systems for Indoor Robots: Design, Applications, and Future Trends in GPS-Denied Environments serves as a valuable resource and inspiration for the reader. Whether you are aiming to build an indoor localization system or simply to understand the state of the art in acoustic positioning, we trust that the knowledge compiled in this book will enrich your efforts. The field of indoor positioning continues to evolve rapidly, and acoustic methods are poised to play an important role in the next generation of indoor navigation solutions. It is our earnest hope that the ideas and techniques presented here will spur further innovation and practical deployments.

Hangzhou, Zhejiang, China
September 2025

Zichen Huang

Contents

Chapter 1
Introduction

1.1 Background and Motivation

1.1.1 The Rise of Indoor Positioning Systems

Modern industries and services increasingly rely on precise location tracking inside buildings, including factories, warehouses, shopping malls, and hospitals. In these indoor environments, global navigation satellite systems like Global Positioning System (GPS) lose reliability or become unusable due to signal attenuation and multipath issues. This creates a need for specialized indoor positioning systems (IPS) that can operate in GPS denied conditions. Such systems enable a range of applications, including asset tracking in manufacturing, guiding autonomous robots in warehouses, and navigation aids for occupants in smart buildings. They form a critical part of the Internet of Things (IoT) infrastructure by providing location context to indoor IoT devices and robots. For example, an IPS can help orchestrate autonomous forklifts on a factory floor or guide shoppers to products in a mall, improving operational efficiency and user experience.

The demand for indoor positioning has grown rapidly over the past two decades. Early research in the 1990s and 2000s explored a variety of techniques, and today, indoor localization has become a vibrant field of both academic research and commercial innovation. Technology giants and startups alike have invested in indoor location-based services for smartphone navigation, asset management, and robotics. As a result, numerous IPS technologies have emerged, each with their own strengths and suited environments. The "rise" of indoor positioning is driven by this broad range of use cases and the limitations of GPS once we move under a roof (Fig. 1.1). The next sections outline why conventional GPS struggles indoors and why certain environments, like greenhouses, present special challenges and opportunities for indoor positioning.

Z. Huang, *Acoustic and Signal-Based Local Positioning*, Navigation: Science and Technology 18, https://doi.org/10.1007/978-981-95-6083-7_1

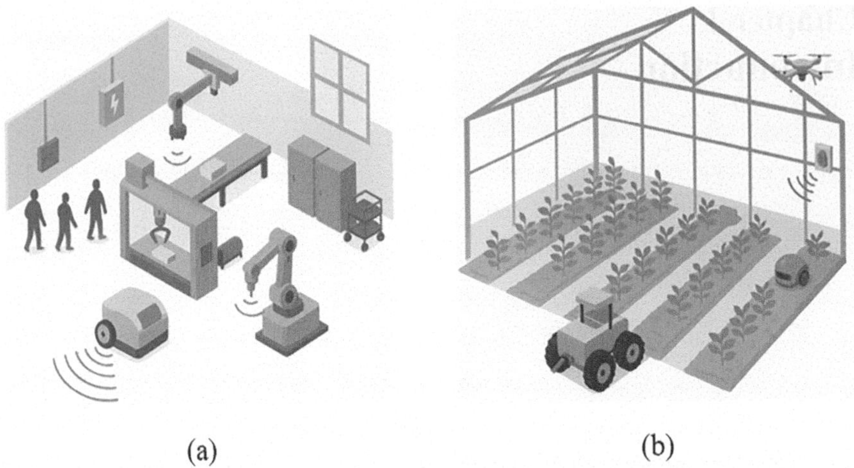

Fig. 1.1 Typical indoor positioning scenarios of the (**a**) industry and (**b**) greenhouse

1.1.2 Challenges in GPS-Denied Environments

Outdoor autonomous machines (like tractors and drones) often rely on GPS, sometimes augmented by Real-Time Kinematic corrections (RTK), to achieve centimeter-level accuracy. For example, agricultural field robots equipped with GPS-RTK and inertial sensors can routinely attain ~2–3 cm positioning accuracy in open sky conditions (Alberto-Rodriguez et al., 2020). However, in indoor or covered environments, the situation is very different. Signals from GPS satellites are weak and high-frequency; they do not easily penetrate building materials and are prone to reflection. Inside a building or a greenhouse, non-line-of-sight transmission, limited penetration capability, severe multipath effects, and weak signal strength cause GPS to fail or degrade unacceptably. In other words, once under a roof (be it metal, glass, or even dense foliage), a receiver may pick up no clear GPS signals at all, or only reflections that confuse its calculations. Even specialized indoor Global Navigation Satellite System (GNSS) repeaters or pseudolite systems struggle to overcome these fundamental challenges of the environment.

The inability to use satellite positioning indoors necessitates alternative solutions. Over the years, many technologies have been adapted or invented to fill this gap, but each must contend with core challenges of indoor radio propagation and logistics. Multipath interference occurs when signals (whether from satellites or indoor transmitters) bounce off walls, ceilings, and equipment, causing the receiver to get delayed, reflected versions in addition to or instead of a direct line-of-sight signal. This is a major source of error in most positioning systems. Signal attenuation is also significant, as walls, floors, and objects absorb and weaken signals in unpredictable ways. These factors mean that an indoor positioning technique often needs more infrastructure (like multiple beacons or sensors) to achieve the same reliability as GPS under open sky. Furthermore, indoor environments are highly diverse, and a solution that works well in a small home or office may not scale effectively to a large factory or a multi-level shopping center. Thus, IPS designers face a complex optimization of accuracy, coverage, infrastructure cost, and robustness to environmental changes. The next subsection introduces the greenhouse as a particularly important use-case environment. It exemplifies many of the challenges faced in GPS-denied settings while also presenting additional unique difficulties.

1.1.3 The Special Case of Greenhouses

Agricultural greenhouses present a compelling and unique case for indoor positioning technology. Greenhouses are essentially large indoor farms, and they are becoming increasingly important for food production. They allow controlled environment agriculture, yielding high-value crops year-round and mitigating the uncertainties of climate. As the global population is projected to reach 9.7 billion by 2050, interest in automated agriculture, including greenhouse robotics, has surged. Greenhouse operations like crop monitoring, precision spraying, harvesting, and logistics can all benefit from automation, but precise positioning of robots is a prerequisite for these tasks. Unlike open fields where GPS-based autoguidance is now common for tractors and combines, greenhouses are GPS-denied environments. Yet they often require even finer precision due to the tight spaces and delicate crops involved.

Several characteristics make greenhouses a special (and demanding) indoor environment for positioning systems. First, the layout of a commercial greenhouse typically consists of long, narrow aisles formed by plant ridges (Fig. 1.2a) or hydroponic benches (Fig. 1.2b). These pathways can be very tight, sometimes only 30 cm wide in ridge cultivation setups, leaving only a few centimeters of clearance on either side of a mobile robot. A positioning error that might be negligible in a warehouse (say 0.2 m) could lead to a collision with plants in a greenhouse. In fact, for

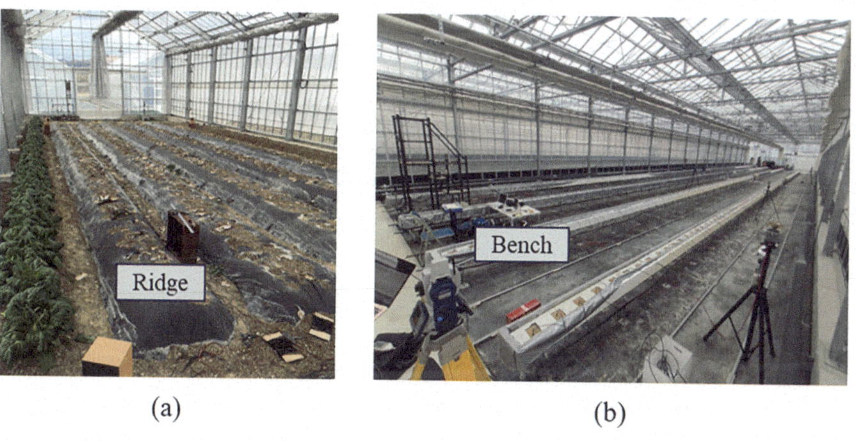

(a) (b)

Fig. 1.2 Typical greenhouse structures of the (**a**) ridge type and (**b**) bench type

some tasks like autonomous ridge farming, a two-dimensional (2D) accuracy of about 5 cm is required to keep the robot centered and avoid damaging crops. In extreme cases (such as precise ridge forming or leveling), straightness requirements can be as low as 2 cm over long runs. These accuracy demands are far stricter than typical indoor navigation needs and even surpass many outdoor farming scenarios.

Second, greenhouses combine the complexities of both industrial indoor spaces and agricultural fields. They are filled with obstacles and occlusions: plants, trellis structures, irrigation pipes, hanging equipment, etc. The environment is dynamic as well—plants grow and can sway, workers and vehicles move around. From a sensing perspective, plants present problematic obstacles. They can absorb or scatter radio signals because of the water content in their leaves, block line-of-sight for optical systems, and at the same time remain irregular in shape and non-static. Third, the microclimate in a greenhouse is harsh on electronics: high humidity, high temperatures, and even water splash or dust are common. Any positioning system deployed there must be robust to these conditions or protected against them. Fourth, the greenhouse interior often lacks distinctive visual features. The endless rows of similar plants create a low-texture environment that can confound vision-based Simultaneous Localization and Mapping (SLAM) systems, which depend on unique visual landmarks. Lighting conditions can also be challenging; while greenhouses have plenty of light in general, direct sunlight and sharp shadows or the spectral conditions under certain films can all hinder machine vision. Lastly, cost and deployment practicality are critical in agriculture. Profit margins in farming can be slim, so any technology must justify its cost. Growers will prefer solutions that do not require extensive retrofitting of the greenhouse or constant maintenance (e.g., it's impractical to place hundreds of QR code markers or reflective targets on the plants). A system that can be installed with minimal disruption, or preferably one that is carried mainly by the robot itself, is highly desirable.

In summary, a greenhouse is a testbed environment that encapsulates many challenges for indoor positioning: no GPS signals, multipath from metal structures and glass, occlusion by dense crops, stringent accuracy and reliability needs, and environmental extremes. At the same time, solving the positioning problem for greenhouses is immensely worthwhile, as it would unlock full automation for an important sector of agriculture. The techniques and insights gained are likely transferable to other indoor domains (such as warehouses or factories with dense shelving, or indoor gardens and vertical farms). The following sections will compare various indoor positioning technologies and explain why this book focuses on a sound-based approach as particularly well-suited for greenhouse applications.

1.2 Indoor Positioning Technologies: A Comparative Overview

Indoor positioning has been tackled with a variety of technologies. Broadly, we can categorize these approaches into radio frequency (RF) systems, vision-based (including SLAM) systems, acoustic (sound-based) systems, and hybrids thereof. Each has its principles, advantages, and disadvantages. This section provides a comparative overview of major indoor positioning technologies, setting the stage for a deeper dive into acoustic positioning in later chapters. We will see that while RF and vision methods are powerful, they also face limitations in environments like greenhouses, which ultimately motivates the choice of a sound-based system.

1.2.1 RF-Based Systems (Wi-Fi, Bluetooth, UWB)

RF positioning is among the most widely used approaches indoors, owing to the ubiquity of radio devices. Common implementations include Wi-Fi access point positioning, Bluetooth Low Energy (BLE) beacon systems, ultra-wideband (UWB) ranging, and even RFID tags. The general idea is to use radio signals emitted by beacons or tags and measure either the signal strength or the signal travel time to infer distance.

Many modern indoor localization solutions leverage existing communication infrastructure. Wi-Fi routers or BLE beacons can double as location anchors. A user's smartphone, for example, can scan for nearby Wi-Fi or Bluetooth signals and use algorithms like fingerprinting (matching observed signal strengths to a pre-measured radio map) or lateration (estimating distance from signal attenuation) to roughly estimate its position. These systems benefit from using devices already present (phones, routers), making them low-cost to deploy. However, their accuracy is limited: typical errors are on the order of 2–5 m for Bluetooth and Wi-Fi in open indoor spaces. Received Signal Strength Indication (RSSI) is notoriously fickle, affected by multipath and human presence, so achieving even 2 m accuracy often requires careful calibration or dense beacon deployment. New developments like Wi-Fi RTT (Round-Trip Time) and Bluetooth 5.1+ angle-of-arrival can improve this to sub-meter accuracy under ideal conditions, but adoption is limited and performance can deteriorate with obstructions. In a greenhouse, Wi-Fi/BLE might serve for zone-level tracking (e.g., which section or row a robot is in), but alone they struggle to meet the centimeter-level accuracy requirement for row navigation.

UWB radio has emerged as one of the most promising high-precision indoor positioning technologies in recent years. UWB devices transmit very short pulses across a broad frequency spectrum (3–10 GHz). By measuring time-of-flight or time-difference-of-arrival between a tag and multiple anchors, UWB can achieve ranging precisions of a few centimeters. In open indoor environments, centimeter-level 2D/3D accuracy is feasible with UWB under line-of-sight conditions.

Commercial UWB systems (e.g., Pozyx or Decawave-based anchors) advertise ~10 cm accuracy in warehouses and factories. Research prototypes have pushed accuracy even further by combining UWB with inertial sensors or applying advanced error correction. These features make UWB a strong candidate for greenhouse positioning. However, UWB is not without challenges. It suffers from multipath and Non Line of Sight (NLOS) issues similar to those of other RF systems. When a direct path is blocked by foliage or metal, the reflected paths can cause significant ranging errors, sometimes amounting to several meters. In a greenhouse with many metal frames and wet leaves, this multipath issue is a primary concern. Mitigation strategies (identifying and ignoring NLOS measurements, using antenna arrays for angle-of-arrival, etc.) add complexity and cost. UWB hardware itself, while dropping in price, is more expensive than Bluetooth or audio hardware. Each robot needs a UWB transceiver, and the infrastructure of anchors must be powered and possibly weather-proofed for the greenhouse environment. Power is another factor: UWB radio can consume significant energy, which might be problematic for battery-operated devices in the field. Summarizing RF methods, UWB offers high accuracy but higher cost and complexity, whereas Wi-Fi/BLE are convenient but relatively coarse in accuracy.

Another RF-based approach uses Radio-Frequency Identification (RFID) tags and readers. RFID tags are very cheap (passive tags cost only a few cents) and require no batteries, which makes them attractive for agricultural settings (for example, one could tag plants, pallets, or locations). For positioning, one strategy is to embed many passive tags in the environment at known coordinates and equip the robot with an RFID reader; as it moves and detects certain tags, it can roughly infer its location. This approach is closer to proximity-based localization. For example, if the robot is detected near tag A15, its position can be inferred as row 5, column 3. The accuracy is limited by tag density, and knowing that the robot is near a given tag may provide a resolution of approximately one meter at best. More advanced RFID localization uses signal phase measurements or multiple antennas to triangulate the tag itself (if the tag is on the robot and fixed readers around). Such systems have achieved sub-meter accuracy in labs, but in practice deploying the required dense infrastructure is challenging. In a greenhouse, hundreds of tags or several specialized readers might be needed, and the metal structure can interfere with radio signals and even detune RFID antennas. Thus, pure RFID solutions are usually better for identification and asset tracking than precision navigation. They could be combined with other methods to provide occasional checkpoints or corrections (for instance, a robot could scan a particular QR code or RFID tag at the end of a row to recalibrate its position).

In summary, RF-based indoor positioning spans a spectrum from low-cost, low-precision methods (BLE beacons, passive tags) to higher-cost, high-precision ones (UWB). Each can be considered for greenhouse deployment, but issues like multipath (especially severe in metal-framed structures) and the need for infrastructure installation in a humid, plant-filled environment must be weighed. These limitations motivate looking at alternative domains like optical and acoustic signals.

1.2.2 Vision and SLAM-Based Approaches

Another major category of indoor localization relies on vision sensors. Cameras, sometimes combined with Light Detection and Ranging (LiDAR) or a depth sensor, enable a robot to perceive its environment and deduce its position. Approaches here often fall under visual SLAM (Simultaneous Localization and Mapping): the robot builds a map of its environment (or uses a known map) and continuously estimates its own pose by recognizing features or landmarks in camera images (Fig. 1.3). Drones and ground robots have successfully used visual SLAM in many settings, leveraging natural features (like corners, edges, textured surfaces) or artificial markers (QR codes, fiducials) for navigation.

Vision-based positioning can achieve high accuracy in structured settings. For example, a LiDAR (laser scanner) SLAM system can easily get accuracy on the order of 1–5 cm in a cluttered indoor environment by matching laser scans to a map. Camera-based systems can also be very precise if there are enough identifiable feature points—algorithms like ORB-SLAM can localize a camera within a few centimeters in a good environment, essentially serving as an indoor "GPS" for a robot. Moreover, vision provides rich information, since a camera can detect obstacles, read signs or text, and perform many tasks beyond simple positioning. For greenhouses, one might imagine using the regular rows of plants as visual features, or placing some colored markers at certain positions.

However, vision and SLAM approaches face significant challenges in greenhouses. As mentioned, the environment often lacks distinct visual features, since rows of identical plants and trellis structures create a highly repetitive scene. This can confuse feature-matching algorithms (many parts of the scene look the same). Lighting can also vary drastically (bright sunlight patches versus shade), which affects camera exposure and the appearance of features. Another issue is that plants

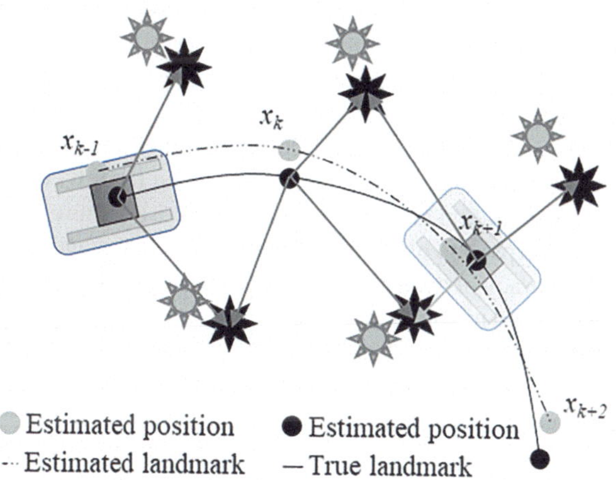

Fig. 1.3 Schematic diagram of SLAM principle

are non-rigid and can move (in the breeze or when touched), violating assumptions of a static map. While LiDAR is more robust to lighting, a 2D LiDAR only captures a slice of the environment (often at a fixed height) which in a greenhouse might be just seeing stems or lower leaves; a 3D LiDAR gives more information but is expensive and generates huge data streams to process. Vision-based methods also require significant computation for image processing and may need frequent recalibration if the environment changes (e.g., growth of plants over weeks could alter the scene).

In practice, purely vision-based localization in a greenhouse tends not to be reliable enough on its own. That said, vision can still play a role in a hybrid system. For example, a robot may use vision to fine-tune its position relative to crop rows or to avoid obstacles that a global positioning system cannot detect. Visual odometry (tracking movement by analyzing consecutive frames) can complement other sensors. But using vision alone as the indoor positioning backbone would demand careful handling of the above issues. In later chapters (especially Chap. 6 on sensor fusion), we will see how visual information might be combined with acoustic positioning to yield greater robustness than either alone. For now, the key point is that vision/SLAM approaches, while powerful in many indoor scenarios, encounter notable difficulties in greenhouse conditions (lack of features, variable lighting, high computation) and thus are not the sole focus of our solution.

1.2.3 Acoustic Positioning: Principles and Potential

Acoustic (sound-based) positioning systems use sound waves, typically ultrasound (above 20 kHz, inaudible to humans) or sometimes audible-range signals, to measure distances for localization. The principle is analogous to the operation of bats or ultrasonic tape measures. By emitting a sound pulse or signal and measuring the time it takes to reach a microphone, the distance can be inferred, since the speed of sound in air is known to be approximately 343 m/s at room temperature. With distances from multiple transmitters (or multiple receivers), the position of a target can be calculated through trilateration. Acoustic positioning has a long history in indoor localization research, dating back at least to the 1990s. One famous example is the Cricket system from MIT, which around 2004 combined ultrasound beacons with RF signals to achieve decimeter-level room positioning (Smith et al., 2004). Many modern systems have built on similar concepts, often using more sophisticated signals (like encoded chirps) and algorithms for better accuracy and noise tolerance.

The potential accuracy of acoustic systems is very high. One fundamental reason is the much slower speed of sound compared to radio. Radio waves travel at approximately 300,000 km/s, whereas sound in air propagates at about 0.343 km/s. This represents a difference of six orders of magnitude. This means that a given absolute timing error translates to a far smaller distance error for sound. For example, a timing uncertainty of 1 μs (1×10^{-6} s) causes an uncertainty of only ~0.34 mm in distance for a sound wave, but that same 1 μs would correspond to ~300 m of uncertainty for a radio-based system (if we tried a time-of-flight measurement with GPS signals, say). In practice, electronics can measure time much more easily to microsecond precision than to nanosecond precision. As a result, achieving, say, 1 cm ranging accuracy is relatively straightforward with audio signals (which have periods in the tens of microseconds and travel at modest speeds) but extremely challenging with RF signals without specialized hardware. Indeed, researchers have demonstrated sub-centimeter accuracy in small-room indoor experiments using ultrasonic systems. In our context, greenhouse acoustic positioning prototypes have already shown the capability for centimeter-level accuracy over tens of meters. For instance, a spread-spectrum acoustic local positioning system (using encoded sound signals to improve robustness) achieved on the order of 20 mm average error in a 30 m × 30 m greenhouse area (Widodo et al., 2014). Another study using a similar approach reported static positioning accuracy around 12 mm in a greenhouse (Tsay et al., 2022). This level of accuracy meets or exceeds the requirements we outlined for even the tightest scenarios (like 5 cm or better for ridge navigation). It essentially offers the promise of "indoor GPS" in terms of precision, which is a key reason to explore sound-based methods for our application.

Beyond accuracy, acoustic systems offer other attractive features, especially when applied in greenhouses.

Acoustic positioning systems are attractive in part because they can be built using low-cost, widely available components. Standard piezoelectric buzzers or small speakers can serve as emitters, while commodity microphones function effectively as receivers. These parts are significantly cheaper than alternatives such as UWB modules or laser-based systems, making the overall setup far more affordable. Many devices, including smartphones, already contain built-in audio hardware that can potentially be repurposed for ultrasonic signaling. In agricultural applications, especially in greenhouses where large-scale deployment is desirable but budgets may be constrained, this affordability becomes a major advantage. A positioning system composed of several fixed speakers and one microphone per robot can be deployed across a greenhouse with modest investment. Furthermore, audio signal processing can often be handled by simple microcontrollers or digital signal processing chips, without the need for highly specialized or expensive computing hardware. The accessibility and scalability of this technology position it as a practical alternative to more costly systems like LiDAR or UWB.

One of the distinguishing features of acoustic systems is that they do not interfere with RF communication networks. Unlike RF-based positioning systems, which operate within congested frequency bands such as 2.4 or 5 GHz, acoustic signals travel as pressure waves through air and occupy a completely different domain. This separation means that an acoustic indoor positioning system can function alongside Wi-Fi, Bluetooth, ZigBee, and other wireless technologies without introducing signal conflict or additional bandwidth demand. In greenhouse environments where wireless sensor networks are increasingly used for environmental monitoring and automated irrigation, this characteristic is particularly valuable. It allows acoustic systems to coexist without degrading communication reliability or requiring additional RF spectrum management. Although acoustic noise sources such as ventilation fans or farm equipment may pose separate challenges, these do not overlap with or degrade RF performance, and vice versa.

Figure 1.4 presents a comparison of various indoor positioning systems, highlighting differences in accuracy, range, and cost. The figure illustrates how systems such as LiDAR, UWB, and acoustic positioning occupy a favourable space in terms of both performance and scalability. These technologies are well-suited for covering large indoor areas, including multi-span greenhouses, through the addition of anchor nodes or signal sources. Other systems, such as RFID, Wi-Fi, and Bluetooth, tend to be more affordable but typically offer lower accuracy or reduced robustness in cluttered environments. When selecting a positioning system for greenhouse applications, it is important to balance multiple criteria including localization accuracy, real-time update capabilities, ease of deployment, and the specific tasks associated with different types of crops.

Among these options, acoustic systems are especially promising for structured but open environments like greenhouses. Despite concerns that sound waves may be easily blocked, acoustic systems exhibit useful propagation behavior in structured but open environments like greenhouses. High-frequency ultrasound has a relatively short wavelength and can partially diffract around small objects or pass through gaps that might impede longer-wavelength radio signals. In greenhouse layouts, where plants and structural elements may obstruct direct paths, careful placement of speakers, especially in overhead positions, can often ensure that each receiver has access to at least one line-of-sight path. Moreover, spread-spectrum techniques, such as encoding signals with pseudo-random sequences or chirps, allow receivers

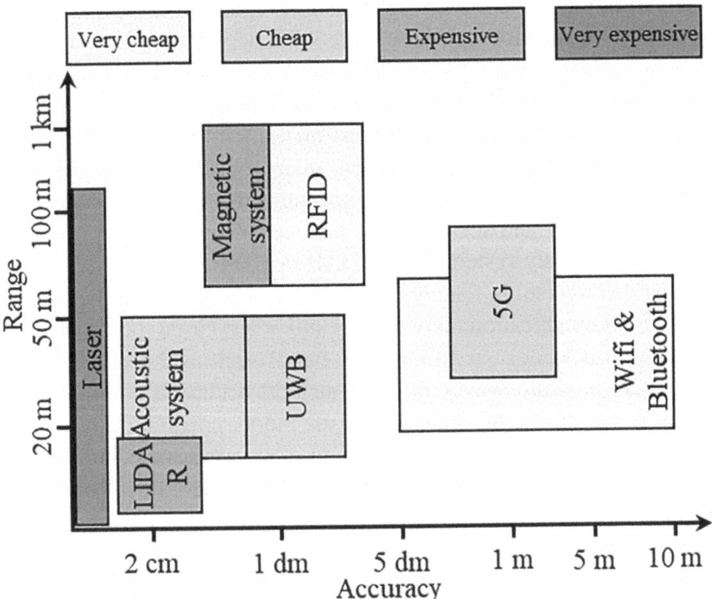

Fig. 1.4 Comparison of indoor positioning systems

to detect the direct signal arrival and distinguish it from later echoes. Because acoustic signals travel more slowly than RF, the delay between the direct and reflected paths is more pronounced, making it easier to isolate the correct arrival time using correlation analysis. Experimental evidence suggests that, when properly designed, acoustic systems maintain reliable accuracy even in moderately obstructed environments (Huang et al., 2020). Multipath effects, which often confuse RF systems, are more manageable in acoustic positioning because of this temporal separation, and signal processing algorithms such as peak detection or threshold filtering can reliably extract the true time-of-flight measurement (Huang et al., 2017).

Given these advantages, sound-based positioning stands out as a promising approach for greenhouse environments. It is capable of high accuracy, uses affordable hardware, and avoids many issues plaguing RF and optical methods. It's no surprise that recent research specifically highlights acoustic IPS as a strong candidate for greenhouses, noting its centimeter-level accuracy and ability to cover entire facilities at low cost. Essentially, an acoustic system can provide indoor localization in a way that mirrors the concept of GPS. Fixed speakers placed throughout the environment act like invisible satellites, each emitting signals that a receiver can use to determine its position. This process takes place entirely within the enclosed structure of a greenhouse, beneath its transparent roof.

1.2.4 Rationale for Choosing Sound-Based Systems

Having reviewed the primary indoor positioning technologies, we now turn to the reasoning behind this book's emphasis on sound-based systems, with a particular focus on greenhouse applications. This choice reflects both the comparative strengths of acoustic methods and the unique demands of the intended environment.

In terms of accuracy and consistency, greenhouse robots often require localization precision within a few centimeters, especially when navigating narrow aisles or performing delicate tasks. Most conventional radio-frequency approaches cannot reliably meet this level of precision, except for ultra-wideband systems, which tend to be expensive and are still vulnerable to signal reflection in enclosed spaces. Vision-based methods, meanwhile, struggle under the fluctuating light and obstructed lines of sight typical in greenhouses. Acoustic systems have shown the ability to meet accuracy demands in such conditions and offer stable performance when supported by suitable signal processing techniques. Prior field studies confirm that acoustic positioning is well suited to deliver the precision and reliability needed in these scenarios.

From a cost and implementation perspective, sound-based systems rely on simple components such as speakers and microphones. This makes them far more affordable than dense arrays of RF anchors or advanced optical systems. Installing a few acoustic beacons across a greenhouse can provide coverage at a fraction of the cost of competing technologies. Maintenance is also straightforward since audio components are easy to clean, replace, or recalibrate, and typically do not require firmware updates or complex synchronization. Furthermore, acoustic systems introduce no additional burden on existing wireless infrastructure. They operate independently of radio-frequency bands, avoiding the risk of interfering with sensor networks already in place for tasks like climate monitoring or irrigation control. In many cases, robotic platforms already include audio hardware, which lowers the marginal cost of integration. The system's design thus supports practical deployment in agricultural settings without significant changes to infrastructure.

The value of acoustic positioning extends beyond greenhouses. Large indoor spaces such as warehouses, underground parking lots, industrial sites, and shopping centers often face similar challenges with GPS denial and the need for accurate localization. The same sound-based strategies developed for greenhouses can be adapted to those environments, making this approach widely applicable. This book focuses on greenhouses not only because of their complexity and relevance to the author's expertise, but also because they serve as a rigorous proving ground for acoustic localization. Lessons learned here are expected to translate well to other domains. While sound-based positioning offers distinct advantages, it is not without limitations. The remainder of this chapter examines the greenhouse as a representative environment, followed by an overview of how the rest of the book explores theoretical foundations, system architecture, common challenges, and potential solutions.

1.3 Greenhouse Environments as a Testbed

Greenhouses provide an excellent testbed for indoor positioning research. They are large enough to pose non-trivial coverage problems (unlike a single room), complex enough to stress the robustness of a system, and yet controlled enough (enclosed, with a known layout) to allow systematic experimentation. Moreover, success in greenhouses has direct real-world impact by enabling agricultural automation. In this section, we examine what makes the greenhouse environment distinctive, what technical requirements arise for robotics operating in it, and how lessons learned in greenhouses can translate to other indoor scenarios.

1.3.1 Environmental Characteristics

A modern greenhouse combines features of both industrial workspaces and living ecosystems, presenting a unique set of environmental characteristics that influence indoor positioning. The structural layout typically includes metal frameworks and transparent walls or roofs made from glass or poly film. Greenhouses vary in size, ranging from compact, single-span tunnels to multi-span complexes spanning several hectares. Internally, the space often follows a grid pattern with long rows of crops separated by narrow walkways. Additional infrastructure such as overhead irrigation rails, heating pipes, and trellises may also be present. These elements create reflective surfaces that affect signal propagation, resulting in multipath behavior for both acoustic and radio signals. In some configurations, long corridors can even channel sound in unintended directions. To maintain positioning accuracy, system designers must carefully select beacon placement to mitigate the effects of these reflections and avoid layouts that amplify echoes through geometric alignment.

Another critical consideration is the high degree of visual and physical clutter inside greenhouses. In contrast to an open warehouse, the space is densely populated with plants and growing equipment that obstruct signals. Leaves and stems can absorb or deflect both radio and acoustic waves, particularly when wet. The level of occlusion is not constant but changes with seasonal crop cycles and plant growth. In addition to the plants themselves, other objects such as gardening tools, pipes, shade cloths, and fans add to the complexity of the environment. These obstructions reduce line-of-sight availability and introduce variable attenuation. One strategy to overcome this challenge is to adjust the frequency of acoustic signals to improve diffraction through foliage. Another is to install transmitters at elevated positions, above the crop canopy, to ensure clearer signal paths.

The internal climate of a greenhouse introduces additional variables. These environments are characterized by high humidity and temperature fluctuations, both of which directly affect sound propagation. While humidity has little impact on radio signals at standard frequencies, it influences the speed and attenuation of sound, particularly at ultrasonic frequencies. The speed of sound increases with temperature and also varies with humidity levels. In practice, temperature and humidity are often unevenly distributed inside a greenhouse, creating vertical and horizontal gradients. These gradients can cause refraction of acoustic waves or introduce measurement errors if left uncorrected. Research has shown that compensating for these variables is essential to maintaining accuracy throughout the day. Moreover, high humidity can enhance the attenuation of ultrasonic signals, especially above 20 kHz. To address this, positioning systems should either incorporate real-time measurements of the local environment or use signal processing methods that are robust to these variations. Fortunately, wind is usually minimal within greenhouses, except near ventilation points, so it poses less concern for acoustic propagation.

Noise is also an important factor to consider. Although greenhouses may seem quiet compared to factories, they contain several sources of background noise. Fans for temperature regulation, mobile carts, irrigation pumps, and small tractors contribute to ambient sound, much of which falls within the audible range. Some mechanical equipment, such as motors and servos, can emit noise into the low ultrasonic range. This presents a challenge for sound-based positioning systems, which depend on maintaining a high signal-to-noise ratio. Acoustic frequencies above 10 kHz tend to be less affected by common greenhouse machinery, making them more suitable for use. Many systems therefore use modulated signals in the 20–40 kHz range to reduce the impact of interference. In scenarios involving drones or aerial robots, additional precautions are needed since these platforms generate broadband noise that extends into the ultrasonic spectrum. Successful deployment depends on identifying the ambient noise profile and selecting appropriate modulation and filtering techniques to preserve system performance. Table 1.1 summarizes key environmental factors, their impact on acoustic signal propagation, and proposed mitigation methods.

Taken together, these factors make greenhouses an ideal test environment for indoor positioning technologies. Their complex combination of reflective surfaces, variable clutter, microclimate instability, and acoustic interference places demanding requirements on any system seeking reliable performance. A solution that can handle the dynamic and challenging conditions inside a greenhouse is likely to perform well in other difficult indoor scenarios. This makes the greenhouse not only a real-world application target but also a proving ground for advancing sound-based indoor positioning.

1.3.2 Technical Requirements for Robotics and Automation

From the perspective of the tasks that greenhouse robots are expected to perform, several technical requirements emerge for an indoor positioning system. One of the most critical is accuracy. Robots working in narrow planting rows may need to maintain a clearance of only a few centimeters, making it essential for the system to achieve positional errors smaller than five centimeters to prevent collisions. While some tasks may tolerate a larger error margin, such as 10–20 cm for basic navigation, operations like precise spraying or fruit picking demand finer accuracy or adaptive corrections. In addition to position, orientation accuracy is also important. Misalignment by more than a few degrees can cause tools to miss their targets. Studies suggest that orientation should be controlled within approximately four degrees to ensure that robots remain properly aligned to crop rows (Dong et al., 2013). This implies that the positioning system must either directly provide orientation information or enable the robot to estimate heading through position tracking or multiple receiver configurations. The key technical specifications needed for positioning systems in greenhouse environments is summarized in Table 1.2.

Table 1.1 Environmental influences on acoustic positioning

Environmental factor	Impact on acoustic positioning	Mitigation strategies
Temperature variations	Alters sound speed and timing accuracy	Real-time compensation, sensors, calibration
High humidity	Increased attenuation at high frequencies	Adjust frequencies, robust algorithms
Multipath (reflections)	Signal interference, false detections	Spread spectrum signals, anchor placement
Acoustic noise	Lower signal-to-noise ratio, interference	Frequency selection, signal modulation
Clutter and occlusions	Signal blocking, reduced line-of-sight	Elevated anchor positions, optimized frequencies

Table 1.2 Technical requirements for greenhouse robots

Technical requirement	Specification	Notes
Positional accuracy	<5 cm	Critical for narrow aisles
Orientation accuracy	~4°	Alignment precision
Coverage area	≥30 × 30 m per zone	Multi-zone coverage possible
Update rate	5–10 Hz (ground robots); 20–50 Hz (drones)	Ensures stable control loops
Latency	<100 ms (ideally ≤50 ms)	Necessary for responsive control
Multi-robot capacity	3–5 robots simultaneously	Requires multiple access solutions
Robustness	Redundancy, fault tolerance	Reliable long-term operation
Deployment ease	Automatic calibration, low intrusion	Simplifies practical implementation

The coverage area is another essential consideration. Greenhouses can be expansive, often extending beyond 100 m in length. The positioning system must provide reliable coverage over the full area in which robots operate. Acoustic systems, particularly those using ultrasound, face range limitations due to signal attenuation, typically functioning well up to 30–50 m. If a single network cannot cover the entire space, the area may need to be divided into zones, each equipped with its own set of anchors. To avoid gaps in localization coverage, systems should be designed to minimize dead zones. Additional anchors or strategically placed transmitters may be necessary in long corridors or at the edges of coverage. Based on earlier field results, maintaining accuracy within a 30 by 30 m area is feasible, and larger greenhouses can be segmented accordingly (Widodo et al., 2014).

Update rate and latency are also important parameters. Most greenhouse ground robots move slowly, typically at speeds between 0.2 and 0.5 m/s, allowing for relatively modest update frequencies. Even one or two position updates per second may suffice for basic motion control at these speeds. However, faster robots or aerial drones may operate at speeds of 1 m/s or more, requiring higher update rates to maintain control and avoid instability. For drones, update frequencies of 20–50 times per second may be necessary. For ground robots, aiming for five to ten updates per second provides a smoother control loop. Achieving high update rates is a challenge for acoustic systems, given the time it takes for sound to propagate and be received. Techniques such as frequency-division and time-division multiplexing can help improve update performance. A practical goal is to maintain latencies below 100 ms, with an ideal target of around 50 ms for responsive feedback.

In situations where multiple robots are active at once, the positioning system must be able to handle concurrent localization without interference. In a typical greenhouse, multiple robots might be performing different tasks in nearby areas. The system must differentiate between them and avoid issues like signal collision or dominance from stronger nearby signals. Acoustic systems are particularly sensitive to what is known as the near-far problem, where a nearby transmitter can overwhelm signals from a more distant one. Solutions to this include the use of unique signal codes, power control, and multiple access schemes. These strategies will be covered in greater detail in later chapters, but the essential requirement is for the system to support at least three to five robots operating in the same zone without noticeable degradation in performance.

Robustness and fault tolerance are also critical for long-term deployment in agricultural environments. Components may fail, become dirty, or be physically disturbed. The system should remain operational even if some anchors are lost, with only minor reductions in accuracy. Temporary obstructions, such as a person walking through the area, should not cause the system to fail. To achieve this, the design must incorporate redundancy and ensure that the loss of one or more nodes does not prevent position estimation. Good system calibration also helps maintain functionality under suboptimal conditions.

Finally, ease of deployment and maintenance is important from a practical standpoint. Farmers and technicians should not need to perform complex setup procedures or measure anchor positions with extreme precision. Systems that support automatic calibration or self-localization of anchors are preferred. Hardware should be installable without interfering with daily operations, and wireless communication or low-intrusion mounts can help achieve this. These practical design concerns, while not directly performance-related, strongly influence real-world usability and adoption.

1.3.3 Relevance to Broader Indoor Scenarios

While our attention is on greenhouses, it's important to highlight that many of the challenges and solutions are applicable to broader indoor scenarios. Large indoor facilities such as warehouses, exhibition halls, airports, and factories share common traits with greenhouses. They are GPS-denied environments, contain obstructions such as shelves, machinery, and crowds of people, and often require precise tracking of assets or vehicles.

For example, consider an indoor warehouse with narrow aisles between tall shelving racks. This layout is analogous in many ways to greenhouse aisles between rows of plants. The need for high accuracy to avoid collisions, the multipath from metal racks, and the potential for occlusion (a forklift in between could block signals) all echo the greenhouse case. An acoustic positioning system proven to work in a greenhouse could be adapted to a warehouse by retuning frequencies (perhaps avoiding loud machinery noise) and recalibrating for the warehouse's temperature profile. Similarly, in a hospital environment where tracking of devices or personnel is required, some facilities already use ultrasound-based systems for asset management. Such systems have clear advantages, since they do not interfere with sensitive medical equipment that relies on radio signals and they can reliably achieve room-level or even bed-level accuracy.

Another scenario is an underground mine or tunnel, which can be regarded as a long indoor space without GPS coverage. Radio communication in such environments can be tricky (RF doesn't propagate well through rock without a relay network), so acoustic or seismic positioning might even be alternatives. The algorithms for signal processing and dealing with multipath in a greenhouse could inform systems for those use cases.

One more domain is smart buildings. As buildings become larger and more complex, such as shopping malls or multi-level industrial plants, a robust indoor positioning system becomes an integral part of the infrastructure. While many current smart building solutions focus on Wi-Fi or Bluetooth due to existing device integration, an acoustic system could complement these, especially for tasks requiring higher accuracy (like guiding robots or carts within a facility). The acoustic approach could also enhance privacy in some cases (sound doesn't penetrate walls well, so localization remains very local and contained, which might be desirable in sensitive environments).

In summary, the greenhouse acts as a microcosm and a stress-test for indoor positioning. If a method can handle the greenhouse, it will likely handle a warehouse, and if it's affordable for agriculture, it will certainly be affordable for commercial buildings. Thus, the work in this book, while centered on greenhouses, aims to contribute knowledge to indoor positioning at large. We will circle back to this broader perspective in Chap. 8 when discussing future directions and potential applications beyond the primary case study.

1.4 Scope and Structure of This Book

Having laid out the motivation and context, we now turn to the scope and structure of this book. The goal of the book is to provide a comprehensive exploration of sound-based indoor positioning systems, using the greenhouse scenario as a unifying example. It is intended for technical professionals and researchers who are interested in indoor localization, robotics in agriculture, or related fields such as sensor networks and automation. We assume the reader has a baseline understanding of engineering principles, but we introduce necessary background on acoustics and signal processing as needed.

1.4.1 Objectives and Summary of Chapters

The objective of this book is to provide a systematic explanation of the theory, design, implementation, and evaluation of a sound-based local positioning system developed for use with greenhouse robots. The content includes fundamental principles such as acoustic wave behavior and time-of-flight ranging, as well as challenges commonly encountered in indoor environments, including Doppler effects and signal reflections. In addition to describing one specific system, the book reviews a range of related studies, including prior work by the authors, to extract practical design insights and broader strategies for effective indoor acoustic positioning. By the conclusion of the book, readers should be equipped with the knowledge needed to design an acoustic positioning system, understand how to overcome its limitations, and evaluate its performance compared with other indoor localization technologies.

This book is intended for an audience of researchers, graduate students, and technical professionals working in areas such as robotics, automation, computer science, and precision agriculture. It is particularly suited to those who are developing or studying indoor navigation systems or alternative positioning technologies. The material may also benefit individuals in the positioning and sensing industries who are seeking specialized solutions for environments like greenhouses. While a basic understanding of signal processing and acoustic or wireless communication principles will be helpful, the book is designed to be largely self-contained, with foundational concepts introduced in the early chapters.

The book does not cover general topics in robotics or agricultural sciences beyond what is necessary to explain the requirements of the positioning system. Although later chapters include experimental validation, this is not a guide to greenhouse management or a general robotics manual. The central theme throughout remains focused on localization technology, with particular emphasis on the acoustic methods that support it. In addition to theoretical explanations and system-level insights, we will provide key MATLAB function code to support practical understanding. These code examples will cover essential components such as time-of-arrival estimation, position estimation, Doppler shift compensation, and other core algorithms relevant to sound-based indoor positioning.

This book is structured to guide the reader from foundational concepts through practical system design and finally to applications of acoustic positioning. Chapter 2 introduces the basic theory of sound propagation and time-of-arrival measurement. It explains how distance can be estimated using sound, and how those measurements contribute to positioning through trilateration. Signal modulation techniques and accuracy considerations are also presented to build the groundwork for later chapters.

Chapter 3 discusses the overall design of an acoustic positioning system. It introduces different architecture options and explains how sound signals are processed from transmission to final position estimation. Strategies such as time division and frequency division are introduced to manage multiple signals, especially for systems involving several robots. Hardware choices and synchronization methods are briefly explained. Chapter 4 explores how movement affects signals, especially through the Doppler effect, and presents compensation methods to correct these distortions.

Chapter 5 focuses on how environmental factors such as temperature, humidity, and noise influence system performance. It offers approaches to mitigate these issues through compensation models and signal design. Chapter 6 looks at combining acoustic data with other sensors like IMUs to increase robustness. It explains how fusion methods can help maintain smooth updates and better orientation tracking. Chapter 7 presents experimental results that validate the system in greenhouse settings, highlighting both performance and practical insights gained during testing. Finally, Chap. 8 reflects on broader applications and future research opportunities, suggesting how this technology could expand beyond greenhouses to other indoor environments.

1.4.2 How to Use This Book

This book can be read in several ways depending on the reader's background and goals. One option is to follow the chapters in order, starting from Chap. 2. This approach is recommended for those new to acoustic positioning, as the material builds progressively. Concepts like cross-correlation introduced early are later used in more advanced topics such as Doppler compensation. Reading sequentially helps avoid gaps in understanding.

Readers who already have experience in indoor positioning or signal processing may choose to focus on specific chapters. Each chapter is written to stand on its own, so it is possible to begin with the section most relevant to one's interest. For instance, someone concerned with practical deployment can start with Chap. 7 on system implementation and then refer back to earlier sections if needed. Those interested in combining sensors might turn to Chap. 6, which includes a summary of the acoustic system before discussing integration. Still, we suggest at least skimming Chaps. 2 and 3, since they define core terms and outline the system used throughout the book.

This book is also intended to serve as a technical reference. Each chapter is divided into sections with clear headings to help readers find specific topics quickly. Readers looking for details like the effect of temperature on sound speed can easily locate the relevant section in Chap. 5. Key references to published research are provided throughout, and the bibliography at the end includes many foundational and recent papers for deeper study. Important equations, definitions, and diagrams are highlighted and often supported by explanatory captions. For those less familiar with signal processing, appendices cover basics such as Fourier transforms and filtering. Readers who prefer detailed analysis will also find mathematical derivations in dedicated sections or notes. Whether used as a tutorial, a guide to practical systems, or a reference manual, the book is designed to be flexible and accessible.

References

Alberto-Rodriguez, A., Neri-Muñoz, M., Ramos-Fernández, J. C., Márquez-Vera, M. A., Ramos-Velasco, L. E., Díaz-Parra, O., & Hernández-Huerta, E. (2020). Review of control on agricultural robot tractors. *International Journal of Combinatorial Optimization Problems and Informatics, 11*(3), 9–20. https://doi.org/10.61467/2007.1558.2020.v11i3.171

Dong, F., Petzold, O., Heinemann, W., & Kasper, R. (2013). Time-optimal guidance control for an agricultural robot with orientation constraints. *Computers and Electronics in Agriculture, 99*, 124–131. https://doi.org/10.1016/j.compag.2013.09.009

Huang, Z., Ono, M., Shiigi, T., Suzuki, T., Harshana, H., Nakanishi, H., & Kondo, N. (2017). Is spread spectrum sound a robust local positioning system for a quadcopter operating in a greenhouse? *Chemical Engineering Transactions, 58*, 829–834. https://doi.org/10.3303/CET1758139

Huang, Z., Jacky, T. L. W., Zhao, X., Fukuda, H., Shiigi, T., Nakanishi, H., Suzuki, T., Ogawa, Y., & Kondo, N. (2020). Position and orientation measurement system using spread spectrum sound for greenhouse robots. *Biosystems Engineering, 198*, 50–62. https://doi.org/10.1016/j.biosystemseng.2020.07.006

Smith, A., Balakrishnan, H., Goraczko, M., & Priyantha, N. (2004). Tracking moving devices with the cricket location system. In *Proceedings of the 2nd international conference on Mobile systems, applications, and services—MobiSYS '04* (p. 190). https://doi.org/10.1145/990064.990088

Tsay, L. W. J., Shiigi, T., Zhao, X., Huang, Z., Shiraga, K., Suzuki, T., Ogawa, Y., & Kondo, N. (2022). Static and dynamic evaluations of acoustic positioning system using TDMA and FDMA for robots operating in a greenhouse. *International Journal of Agricultural and Biological Engineering, 15*(5), 5. https://doi.org/10.25165/j.ijabe.20221505.6796

Widodo, S., Shiigi, T., Than, N. M., Kikuchi, H., Yanagida, K., Nakatsuchi, Y., Ogawa, Y., & Kondo, N. (2014). Wind compensation for an open field spread spectrum sound-based positioning system using a base station configuration. *Engineering in Agriculture, Environment and Food, 7*(3), 3. https://doi.org/10.1016/j.eaef.2014.04.001

Chapter 2
Fundamentals of Acoustic Positioning

Indoor acoustic positioning systems operate by transmitting and receiving sound signals to estimate the position of objects or mobile platforms within enclosed environments. In contrast to outdoor scenarios where satellite-based technologies such as GPS are highly effective, indoor environments present a range of challenges, including signal blockage, multipath propagation, and acoustic reflections. These complexities require alternative positioning solutions that rely on sound, particularly in situations where electromagnetic interference must be avoided or where high spatial resolution is necessary with minimal infrastructure. This chapter introduces the foundational concepts of acoustic positioning, beginning with a discussion of how sound propagates in indoor environments. It then outlines key measurement techniques that utilize time, angle, and amplitude information to determine position. Understanding these physical and algorithmic principles is essential for the development of accurate and reliable indoor localization systems, applicable in a variety of scenarios such as industrial buildings, greenhouses, and autonomous robotic navigation.

2.1 Acoustic Signal Propagation

Acoustic wave propagation in indoor environments is governed by many of the same principles as general acoustics, but with specific considerations for enclosed spaces and obstacles. Key phenomena include attenuation (energy loss with distance and medium), reflection (bouncing off surfaces), absorption (conversion of sound energy to heat in materials), refraction (bending of sound due to medium changes), and diffraction (bending around obstacles). These effects collectively

Supplementary Information The online version contains supplementary material available at https://doi.org/10.1007/978-981-95-6083-7_2.

Z. Huang, *Acoustic and Signal-Based Local Positioning*, Navigation: Science and Technology 18, https://doi.org/10.1007/978-981-95-6083-7_2

determine how an acoustic signal travels from a source to a receiver in a room or indoor space. Understanding these principles is crucial for designing indoor acoustic positioning systems, as they affect signal strength, clarity, and timing.

In free-field conditions (open space with no reflections), sound spreads spherically and its intensity drops following the inverse-square law (Fig. 2.1). In decibel terms, a point source's sound pressure level (SPL) decreases approximately as:

$$Lp(d_2) = Lp(d_1) - 20\log_{10}\left(\frac{d_2}{d_1}\right) \tag{2.1}$$

where, $Lp(d_1)$ is the known sound pressure level at the first location, $Lp(d_2)$ is the unknown sound pressure level at the second location, d_1 is the distance from the noise source to location of known sound pressure level, and d_2 is the distance from noise source to the second location.

In practice, indoor environments are not fully free-field, but this formula provides a baseline for geometric spreading loss. Atmospheric absorption can add further attenuation, especially at high frequencies and long distances. For instance, high-frequency components may dissipate in air (humidity and air composition cause frequency-dependent absorption), so distant sounds often lose treble content. In typical indoor distances (a few meters in indoor built environment or 50 m greenhouse environment), air absorption is usually negligible, but for ultrasonic signals (20–40 kHz) even moderate distances can introduce noticeable attenuation due to air viscosity and thermal losses.

When sound waves encounter surfaces (walls, ceiling, floor, objects), a portion of the wave reflects back into the room and the rest may be absorbed or transmitted through the material. The ratio of reflected to incident energy is characterized by the reflection coefficient (and the complementary absorption coefficient for absorbed energy). Hard, dense materials like concrete or glass have high reflection coefficients (close to 1.0), meaning most acoustic energy reflects and very little is absorbed. Soft or porous materials (carpets, acoustic foam, curtains) have lower reflection coefficients and higher absorption—they soak up sound and convert it to slight heat, thereby damping the wave. For instance, a concrete or brick wall might reflect ~95% of incident acoustic energy (absorption coefficient ~0.05), whereas a heavy curtain could absorb 50% or more (absorption coefficient ~0.5) at certain frequencies. In enclosed spaces, multiple reflections off room boundaries lead to complex propagation paths (see Multipath and Reverberation below). In some specialized indoor

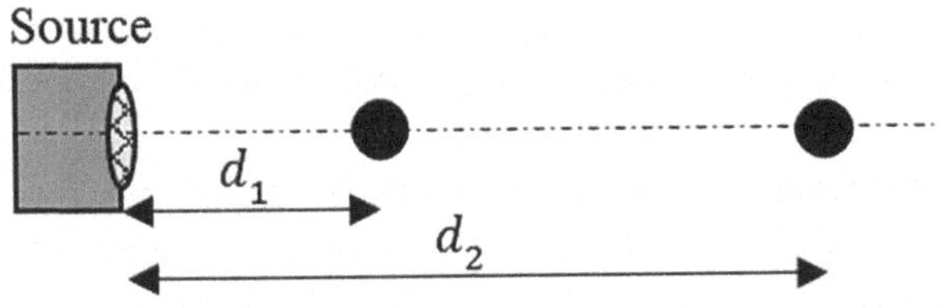

Fig. 2.1 Attenuation with distance

environments like greenhouses (or indoor farms), the surfaces and contents can markedly affect reflection/absorption. Greenhouses often have large flat surfaces (glass or polycarbonate walls) which strongly reflect sound, but also contain lots of plants which can absorb sound. In a greenhouse environment an ultrasonic positioning signal would not only reflect off rigid surfaces (glass frames, walls) with high efficiency, but also be partly absorbed and scattered by plants, effectively reducing signal strength and limiting the distance it can travel without considerable loss.

Refraction is the bending of sound waves due to spatial variations in sound speed. Indoors, this can occur if there are temperature gradients or the sound passes through different media. A common example is sound traveling from indoor air through an open doorway to another room or hallway. The change in medium, since air in one room may have a slightly different temperature or humidity than in the next, can bend the path of the wave. In large open indoor spaces, temperature gradients (warm air near the ceiling from heating, cooler near the floor) can cause slight refraction of sound upward or downward. For instance, on a warm day in a large atrium, the air near the ceiling may be warmer and the speed of sound slightly higher compared with the air near the floor. As a result, a sound wave traveling upward can refract toward the cooler and slower medium near the floor. In outdoor environments, this effect is even more pronounced. At night, cooler air close to the ground can bend sound downward, thereby increasing its range, whereas hot air near the ground during the day can bend sound upward and create acoustic shadows. Indoors the scale is smaller, but the principle remains the same. Any spatial variation in the properties of the medium can deflect acoustic paths. For acoustic positioning, refraction is usually a minor concern compared to reflections, but it could subtly affect long-range measurements or when sound passes through ventilation shafts, open windows, etc., where the medium changes.

Diffraction is the ability of sound waves to bend around obstacles or spread after passing through openings. It is more pronounced when the wavelength is large relative to the obstacle or aperture. Low-frequency sound, with a wavelength on the order of meters, diffracts strongly around furniture, partitions, or doorways. It can wrap around corners and fill shadow zones. High-frequency sound, with shorter wavelengths such as ultrasounds in the millimeter range, diffracts much less and behaves almost like light beams. It casts sharp shadows behind obstacles. In an indoor positioning context, audible-frequency signals (say 2–10 kHz) can bend around obstacles to some extent, providing coverage even behind obstructions, whereas ultrasonic signals (~20–40 kHz) tend to require line-of-sight or will be drastically weakened if obstructed (Filonenko et al., 2013). This situation represents a trade-off. Higher frequencies allow greater precision, and in the case of ultrasound they are inaudible to humans, but they are also more susceptible to occlusions. Lower frequencies travel further and around obstacles better, but provide less precision and may be audible. Diffraction also contributes to frequency-dependent multipath. For example, a doorway can act as a secondary source through aperture diffraction, spreading sound into an adjacent space.

In an enclosed room, a receiver typically does not only get the direct line-of-sight sound from the source; it also receives reflected copies of the sound from various surfaces (Fig. 2.2). These multiple paths cause interference, which can add or cancel

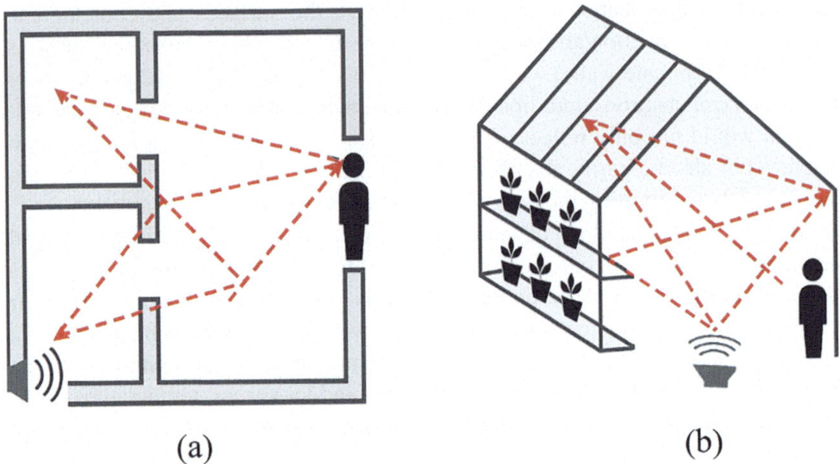

Fig. 2.2 The multiple paths in (**a**) an indoor built environment and (**b**) a greenhouse environment

out depending on their relative phase. This is analogous to multipath in radio, where constructive interference occurs at some points and destructive interference at others, producing a spatial pattern of peaks and nulls in signal strength that is known as fading. For an acoustic positioning system, multipath is problematic because it can distort the signal timing and amplitude. For example, if you send a short ultrasonic pulse, the receiver might hear a primary pulse, then a series of echo pulses (reflections) arriving slightly later. These echoes (multipath) can confuse time-of-flight measurements. Multipath can also cause frequency-selective fading, in which certain frequencies are reinforced while others are canceled, thereby altering the spectrum of the signal. In short, multipath in indoor acoustics can blur the temporal signal and create frequency-response irregularities. The ultrasonic positioning in rooms must account for echoes and often the first arriving (shortest path) signal is used and later echoes are filtered or ignored. Signal processing strategies such as correlation receivers and pulse shaping can mitigate some multipath effects by distinguishing the direct path from delayed ones, or by using techniques like time gating (listening only in the expected time window for direct arrival) (Yan et al., 2023). Nonetheless, multipath remains a major challenge for acoustic indoor positioning because it can mimic or mask the true propagation delay.

Reverberation is an extreme case of multipath where so many reflections occur that the sound persists for a time even after the source stops. It is characterized by a reverberation time (RT60), which is defined as the time required for the sound to decay by 60 dB after the source ceases. Small, absorbent rooms have short reverberation times (perhaps <0.5 s), while large halls with hard surfaces have long reverberation (several seconds). In a reverberant space, an acoustic positioning signal (especially if it's something like a tone or chirp) will generate a cloud of overlapping echoes. This can severely degrade clarity. Reverberated sounds "stack up," adding a persistent ambient tail that can mask new sounds. The difference between a single clear echo and reverberation is that in reverberation, reflections arrive from all directions continuously, creating a dense, continuous decay rather than discrete echoes. For indoor robot positioning, reverberation means that a signal's energy from previous transmissions may still be bouncing around when the next signal is emitted, potentially causing interference. Reducing reverberation can be done by adding absorbing materials to the environment or by using signals that can be distinguished even in reverberation (again, spread-spectrum techniques or pulse compression methods are helpful). Some advanced localization algorithms even exploit reverberation by treating the room reflections as additional information, but generally, reverberation is a hindrance that reduces positional accuracy.

In summary, indoor acoustic propagation is influenced by fundamental wave behaviors and various degradation factors. A successful indoor acoustic positioning system must be designed with these factors in mind. It should ensure adequate source levels to overcome attenuation and noise and achieve a high signal-to-noise ratio, employ signal processing techniques to handle multipath and reverberation, operate at frequencies that balance diffraction for coverage with resolution, and leverage propagation models to calibrate distance measurements.

2.2 Spread Spectrum Techniques

To improve the robustness of acoustic positioning signals in noisy, echo-prone indoor environments, spread spectrum techniques can be employed. Spread spectrum is a family of signal modulation methods originally developed for secure and reliable radio communication, but the underlying concepts are equally applicable to acoustic signals. The core idea is to spread the signal's energy over a much wider bandwidth than what is actually needed for the information being sent. This approach yields several benefits that are highly relevant to indoor acoustic systems. It provides resistance to interference and noise, enables the distinction of signals from multipath echoes, and offers a degree of inherent security or privacy, since a spread signal is difficult to detect or jam without knowledge of the spreading key.

In the context of indoor acoustic positioning, spread spectrum can be used to encode the timing or identity of acoustic "pings" in a robust way. For example, instead of emitting a single-frequency tone pulse (which a loud noise or an echo could easily disturb), the system could emit a spread-spectrum chirp or coded signal that can be reliably picked out of background noise via correlation techniques. The main spread spectrum approaches are Direct Sequence Spread Spectrum (DSSS), Frequency Hopping Spread Spectrum (FHSS), and Chirp Spread Spectrum (CSS).

In DSSS, a high-rate pseudo-random code sequence is used to directly modulate the transmitted signal, spreading a narrowband data signal into a much wider band. Essentially, the data are multiplied by a pseudo-noise (PN) code, which is a sequence of +1 and −1 chips that rapidly flip the phase of the signal (Fig. 2.3). The result is a noise-like wideband signal. Only a receiver that knows the same PN code can correlate with the received signal to "de-spread" it, recovering the original data. All

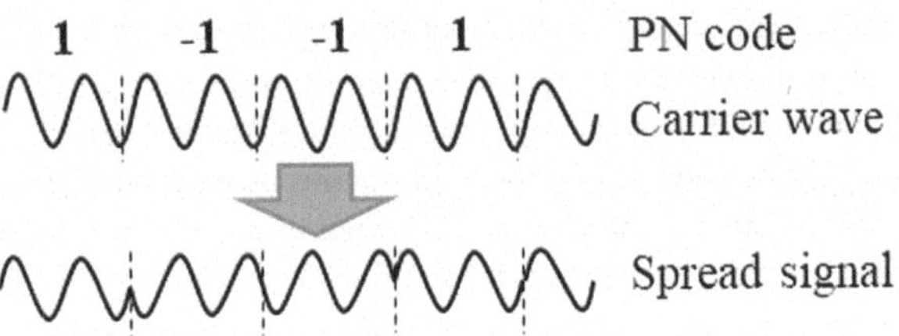

Fig. 2.3 Example of direct sequence spread spectrum (DSSS)

other receivers (or noises) that don't know the code see only a low-power noise-like signal.

The key advantage of DSSS is that it provides a processing gain equal to the spread factor (the ratio of chip rate to data rate). A DSSS acoustic system could allow a very weak ultrasonic chirp to be detected in a noisy factory because the correlation process "boosts" the SNR by the spreading factor. Another advantage of DSSS is interference rejection. Narrowband noises (e.g. a tonal hum at 5 kHz from a machine) will be spread across the large bandwidth at the despreading stage, effectively diluting their power. DSSS is especially effective against narrowband jammers or interference, as the despreading correlator treats any signal not following the exact PN code as noise and spreads it out. In addition, multiple DSSS signals can coexist, if each device/anchor uses a different orthogonal or low-cross-correlation code, they can transmit simultaneously on the same frequency band with minimal mutual interference.

FHSS takes a different approach to spreading. Instead of distributing the signal continuously over a wide band, it rapidly hops the carrier frequency through a sequence of channels. The transmitter and receiver agree on a pseudorandom hopping pattern. During a brief interval known as the dwell time, the carrier remains at frequency f_1 while carrying the data, which may be a tone, frequency shift keying, or another modulation scheme. In the next interval it moves to frequency f_2, and the sequence continues accordingly. Over time, the signal's energy is distributed across a wide range of frequencies, but at any given instant it is narrowband on one channel. The sequence of hops is the key—only a receiver that knows the sequence (and timing) can follow the transmitter and demodulate the message.

The key advantage of FHSS is that it is very effective against narrowband interference and frequency-selective fading. If one channel is noisy or blocked, the next hop is likely to be clear, so the system only loses a small fraction of the transmission. By spending little time on each frequency, FHSS "evades" interference. For example, if a persistent interferer is present at 40 kHz in an acoustic system, an FHSS transmitter may occupy 40 kHz for only about 10 ms out of each second. Any data lost during that hop can often be recovered through error-correction, while for the remainder of the time the system operates without interference on other frequencies. FHSS also provides a level of security through obscurity. An unintended receiver would need to scan a broad band to capture the signal, and if the timing is missed, the transmission appears as short bursts at random frequencies. Other receivers not synchronized will "only register a slight increase in the overall noise level" when an FHSS signal is present.

Chirp Spread Spectrum uses chirp signals (frequency-swept tones) for spreading. A chirp is a signal whose frequency continuously increases or decreases over time. For example, an up-chirp might sweep from 20 to 30 kHz over a duration of 10 ms. Chirp signals inherently occupy a broad bandwidth (equal to the sweep range) and have a time-varying frequency. In CSS, data can be encoded in the initial phase or frequency of the chirp, or by using different chirp rates, etc., but the important part is that the entire chirp bandwidth carries the information. At the receiver, a matched filter or correlator for that chirp can be used to detect the signal with high processing gain.

Chirp signals have excellent correlation properties. A linear chirp exhibits a narrow and strong autocorrelation after de-chirping, which allows reliable detection even under low signal-to-noise ratio conditions. Because the chirp's energy is spread over time and frequency, it is resistant to both narrowband noise and multipath. In fact, one major advantage is that chirps are resistant to multipath fading and can even be designed to be partially resistant to Doppler effects. When a chirp propagates in a multipath environment, the reflections appear as time-shifted copies. If the delay is not too large, the matched filter can still capture most of the linearly increasing frequency content. Moreover, since the frequency is changing, if at some moment a certain frequency is in a fade, the chirp quickly moves out of that fade in frequency. This is different from a steady tone that could be nulled out continuously by a standing wave. The wideband nature of chirps makes them robust to frequency-dependent nulls and interference, similar to DSSS in effect (using the whole band). Another benefit is that chirps, especially those using linear frequency modulation (LFM), are well known in radar and sonar for their pulse compression capability. A long chirp can be processed and compressed into a very sharp time-domain peak, which improves range resolution without requiring a high instantaneous amplitude.

In the context of acoustic positioning systems for indoor robots, selecting an appropriate spread spectrum technique is crucial to ensure accurate localization amidst challenges such as multipath propagation, ambient noise, and Doppler effects. The three primary spread spectrum methods, DSSS, FHSS, and CSS, each offer distinct advantages and limitations. The following comparative analysis focuses on their applicability to acoustic positioning systems (Table 2.1).

2.3 Positioning Methods

Indoor acoustic positioning systems primarily employ four fundamental techniques to estimate distances or directions: Time of Arrival (ToA) (Fig. 2.4a), Time Difference of Arrival (TDoA) (Fig. 2.4b), Angle of Arrival (AoA) (Fig. 2.4c), and RSSI. Each method relies on a different physical signal metric, such as propagation time, differential time, incoming angle, or signal attenuation, and therefore each has unique implementation requirements and challenges. These techniques have been

Table 2.1 Comparison between the spread spectrum approaches

Feature	DSSS	FHSS	CSS
Signal type	Pseudo-random noise (PN) sequence	Rapid frequency hopping	Frequency-modulated chirp signals
Multipath resistance	Moderate; susceptible to multipath interference	High; frequency hopping mitigates multipath effects	Excellent; inherently resistant to multipath and Doppler effects
Synchronization	Requires precise timing for PN sequence alignment	Requires synchronization of hopping patterns	Simpler synchronization due to linear frequency variation
Data rate	High; suitable for applications requiring rapid data transmission	Moderate; lower than DSSS	Low to moderate; optimized for robustness over data rate
Power consumption	Moderate	Low; efficient due to narrowband transmission at any given time	Low; designed for energy-efficient applications
Implementation	Complex; needs precise timing mechanisms	Simpler hardware requirements	Moderate; requires specific algorithms for chirp signal processing
Best use case	Environments requiring high data rates with moderate interference levels	Environments with high interference and need for robustness	Dynamic indoor environments requiring energy efficiency and robustness

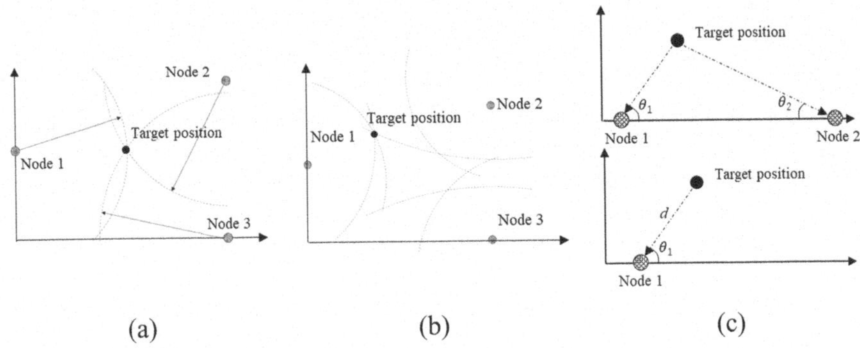

Fig. 2.4 Positioning algorithms of (**a**) ToA, (**b**) TDoA, and (**c**) AoA

widely studied (even in radio frequency contexts) and adapted to acoustic signals such as ultrasonic beacons for indoor robot localization. In general, time- and angle-based methods (ToA, TDoA, AoA) geometrically compute the position from measured distances or angles, offering high accuracy but being sensitive to multipath and NLOS propagation. By contrast, RSSI-based methods infer range from signal strength and can be simpler to deploy, though they suffer from environmental attenuation effects. This section discusses each technique in detail, including their core principles, mathematical formulation, practical challenges in indoor acoustic environments, advantages and limitations for sound-based systems, real-world usage examples (particularly with ultrasonic signals), and common mitigation strategies or hybrid integrations to improve performance.

2.3.1 *Time of Arrival (ToA)*

The ToA approach determines distance by measuring the absolute travel time of an acoustic signal from a transmitter to a receiver. Given a known signal emission time, the propagation delay t_i between transmission and reception can be converted to range using the speed of sound v_i. In its simplest form, the distance d_i is obtained as:

$$d_i = v_i \times t_i \tag{2.2}$$

where, v_i is the estimated velocity of distance i, and t_i is the ToA of distance i. For example, if an ultrasonic pulse is emitted and later detected after 20 ms, and assuming v_i is 343 m/s (sound speed in air at room temperature), the path length d_i can be calculated as 6.86 m (=343 × 0.020).

Meanwhile, distance i also can be calculated using target position (x_0, y_0, z_0) and known position of the node i (x_i, y_i, z_i).

$$d_i = v_i \times t_i = \sqrt{(x_i - x)^2 + (y_i - y)^2 + (z_i - z)^2} \tag{2.3}$$

The unknown target position (x, y, z) can be estimated using techniques such as Taylor-series expansion and the least squares algorithm, which iteratively refine the estimated position based on measured distances. Figure 1.3a illustrates the principle of the ToA positioning method, where each anchor node serves as the center of a sphere, and the radius corresponds to the measured distance calculated from Eq. (2.2). The intersection point of these spheres, indicated by the overlapping region (red point), represents the estimated target position. Thus, at least three distance measurements from distinct anchor nodes are required to accurately localize the target.

The accuracy of positioning systems based on ToA is primarily influenced by ranging precision, accurate estimation of propagation velocity, and geometric factors collectively termed dilution of precision (DOP). The theoretical resolution of a ToA-based positioning system can be determined using the following equation:

$$\text{Resolution} = \frac{v_i}{f_s} \tag{2.4}$$

where, f_s is the sampling frequency.

A critical requirement for one-way ToA ranging is time synchronization between the transmitter and receiver. The receiver must know exactly when the acoustic signal was sent in order to measure propagation time. Any clock offset or uncertainty in emission time directly adds to range error. Because the speed of sound is relatively slow, even a small timing error can lead to a significant distance error. For example, a timing offset of 1 ms corresponds to an error of about 34 cm. In indoor scenarios, achieving nanosecond or microsecond-level synchronization (as is needed for centimeter accuracy) is non-trivial. Some systems address this by using an auxiliary radio frequency (RF) signal or a wired trigger to notify the receiver of the transmitting time. For example, a transmitter might send a radio message at the moment of emitting an ultrasonic pulse; the receiver uses the radio signal (which travels effectively instantaneously in comparison) to start a timer and then measures the arrival time of the slower ultrasound, thereby computing the time-of-flight. This RF–ultrasound synchronization technique converts an absolute ToA measurement into a TDoA measurement (between RF and acoustic signals) to eliminate the need for shared clocks, as famously implemented in the MIT Cricket indoor location system (Priyantha, 2005).

When line-of-sight is available and synchronization issues are managed, ToA can yield very high accuracy. Because sound travels much slower than radio waves, measuring time-of-flight in the millisecond range is feasible with inexpensive microcontrollers or sound cards, and the achievable distance resolution can be on the order of a few millimeters given microsecond timing precision. Indeed, indoor ultrasonic systems have demonstrated ranging accuracies on the order of centimeters. A notable example is the Spread Spectrum sound-based local positioning system (developed at Kyoto University), which used the receiver on the robot to pick up acoustic signals from emitters in a greenhouse. The accuracy and relatively straightforward geometric foundation of ToA are strong advantages. Additionally, acoustic ToA systems can leverage the existing audio hardware (speakers and microphones) on robots or smartphones, making them cost-effective in terms of hardware. However, the reliance of ToA on time synchronization is a major limitation. Without an aiding mechanism, such as an RF link or a two-way ranging protocol, it can be impractical to maintain perfectly synchronized clocks between a mobile robot and fixed beacons.

In summary, ToA is a powerful method for acoustic positioning with proven high accuracy, but it requires careful system design (synchronization aids, robust signal processing, and possibly augmentation with other sensors) to perform reliably in complex indoor spaces.

2.3.2 Time Difference of Arrival (TDoA)

The TDoA localization also utilizes propagation time, but in a relative manner that can eliminate certain synchronization needs. Instead of measuring an absolute travel time, TDoA systems measure the difference in arrival times of the same signal at multiple receivers or the difference between multiple signals arriving at one receiver. The key idea is that for two anchors (receiving stations) at known locations, the locus of points that yields a constant time difference Δt between arrivals at those two anchors is a hyperbola (in 2D) with the two anchors at its foci. Geometrically, if d_i and d_k are the distances from the target to anchor i and k respectively, their difference is directly proportional to the arrival time difference:

$$\Delta t \cdot v = (t_i - t_k)v = \sqrt{(x_i - x)^2 + (y_i - y)^2 + (z_i - z)^2} - \sqrt{(x_k - x)^2 + (y_k - y)^2 + (z_k - z)^2} \tag{2.5}$$

where, t_i and t_k represent the arrival times at anchors i and k respectively, v is the signal propagation speed, and (x, y, z) denotes the unknown position of the transmitting device. Each pair of anchors provides an equation of this form, and with at least three anchors one can solve for an intersection point (the target's position) that satisfies multiple hyperbolic constraints. Typically, one anchor is designated as a time reference. For instance, the time differences can be measured between each anchor and the reference anchor, producing two or more hyperbolas that intersect at the target location.

A significant advantage of the TDoA approach is its reduced synchronization requirement compared to the ToA method. In TDoA-based systems, precise knowledge of the transmitter's send time is unnecessary. This flexibility simplifies the system design, either by allowing the transmitter (often a mobile node) to emit pulses without its own synchronized clock, or by enabling a mobile receiver to only measure relative arrival times from multiple synchronized transmitters. Specifically, in a configuration where multiple receivers (anchors) detect signals from a single mobile transmitter, synchronization is limited to the receivers, removing the need for clock synchronization at the transmitter. Conversely, in a multi-transmitter scenario, anchors emit signals in a coordinated sequence controlled by a master clock, allowing the receiver to deduce position simply by observing relative timing differences, without needing synchronization to the anchors' clocks.

However, TDoA does not entirely eliminate multipath issues, as reflected signals can distort measured arrival-time differences. Precise synchronization among receiving anchors, meticulous calibration of anchor positions, and optimized geometric anchor placement via GDOP analysis (Zhao et al., 2022) remain essential to maintain high localization accuracy in indoor acoustic environments.

2.3.3 *Angle of Arrival (AoA)*

The AoA positioning methods determine the location of a target by measuring the direction from which acoustic signals arrive at an anchor point, rather than directly computing distances.

Figure 2.5 illustrates an example of angle estimation used in AoA-based positioning systems. As depicted, acoustic waves emitted from the speaker reach the two microphone elements with a slight time difference due to their spatial

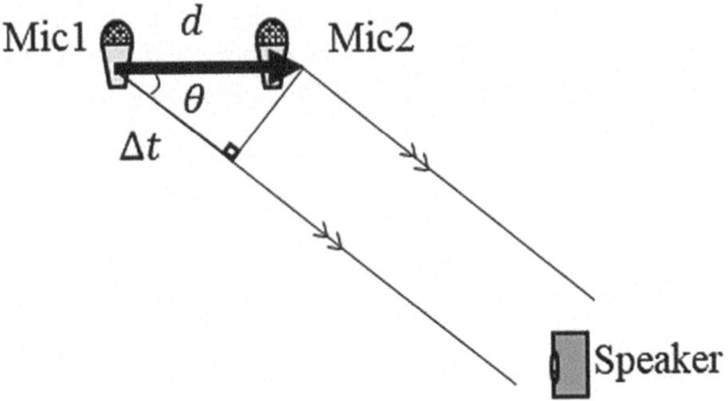

Fig. 2.5 A example of angle measurement using two microphones

separation (baseline d). By precisely measuring this time difference Δt, the angle θ at which the acoustic wavefront arrives relative to the perpendicular orientation (broadside direction) of the microphone array can be accurately determined. The mathematical relationship for calculating the incidence angle is given by:

$$\Delta t = \frac{d}{v} cos\theta \tag{2.6}$$

where, v is the speed of sound in the medium (approximately 343 m/s in air at room temperature), and θ is the angle between the arriving wavefront and the perpendicular (broadside) direction of the microphone array. Solving for θ, we obtain:

$$\theta \approx \arccos\left(\frac{v \cdot \Delta t}{d}\right) \tag{2.7}$$

This simple two-microphone configuration allows estimation of the incoming angle in the plane of the array. For more precise or three-dimensional (3D) direction estimation, larger linear or planar arrays may be used. In such cases, beamforming or high-resolution spectral estimation algorithms, such as MUSIC (Multiple Signal Classification) or ESPRIT (Estimation of Signal Parameters via Rotational Invariance Techniques), are often applied. These algorithms exploit phase and time delay patterns across multiple array elements to determine the signal's direction with high resolution and robustness to noise.

AoA offers several advantages. It does not require time synchronization between transmitter and receiver, simplifying system design compared to ToA or TDoA. It also requires only one acoustic transmission to estimate direction, reducing signal traffic and conserving energy. AoA systems using ultrasonic frequencies and microphone arrays with baselines of tens of centimeters can achieve sub-meter accuracy, making them suitable for indoor applications like robotic navigation and docking. However, AoA also faces challenges. Accurate angle estimation requires precise array calibration; even small errors in microphone placement or timing can introduce significant angular inaccuracies. Moreover, multipath propagation, which is common in indoor environments, can cause reflected signals to arrive from misleading directions and thereby confuse the estimation algorithm. In NLoS conditions, AoA may misinterpret a reflection as the true direction to the source.

To mitigate these effects, systems often use wideband signals, directional beamforming, or filtering techniques to isolate the earliest (likely direct) arrival. Despite these measures, AoA accuracy decreases with distance: a small angular error results in a larger position error the farther the target is. Additionally, at least two angle measurements are needed to resolve 2D position, limiting AoA's standalone use in some settings.

In summary, AoA is a robust and efficient method for indoor acoustic positioning, particularly when directional guidance is important. While its performance can be limited by multipath and calibration errors, AoA becomes highly effective when used in hybrid systems or combined with range-based methods like ToA or RSSI.

2.3.4 *Received Signal Strength Indication (RSSI)*

The RSSI refers to using the amplitude (or energy) of the received signal as a proxy for distance. As a sound wave travels through air, it undergoes geometric spreading and energy loss. In an ideal free-space scenario (no obstacles or reflections), acoustic intensity follows an inverse-square law. Thus, the farther the receiver is from the transmitter, the weaker the signal it detects. By measuring the received signal strength and knowing the transmit power, one can estimate the range.

In practice, an acoustic transmitter might emit a known-amplitude tone or a calibrated sound pattern, and the robot's microphone measures the amplitude to infer how far away it is. If multiple transmitters or beacons are within range, the robot can perform triangulation by converting each RSSI measurement into a rough distance and then applying multilateration similar to ToA. Each estimated distance defines a circle around a beacon, and the intersection of these circles provides the position estimate. Another approach is fingerprinting: rather than using a theoretical model, the system can be trained by measuring RSSI at many known locations, building an acoustic-map of signal strength. Then, when the robot is at an unknown position, it compares the current RSSI readings (from one or multiple transmitters) to the database to find the closest match, effectively interpolating its position. This technique has been extensively used in Wi-Fi localization and could analogously be applied to acoustic signals if one had a static ambient sound or beacon signals in the environment.

One of the main advantages of RSSI-based systems is their ability to function in NLoS conditions, where ToA methods might fail due to the absence of a direct path. In such scenarios, RSSI can still provide degraded yet usable measurements, offering a form of redundancy in challenging environments. Additionally, RSSI measurements can be integrated with other localization techniques, such as ToA or AoA, to enhance overall system robustness, especially in environments where certain measurements are unreliable.

However, RSSI-based acoustic positioning faces significant challenges, particularly in indoor environments. Multipath propagation, where signals reflect off surfaces and arrive at the receiver via multiple paths, can cause constructive and destructive interference. This results in rapid fluctuations in signal strength over small distances, undermining the assumption of a smooth inverse-square relationship between distance and signal strength. Environmental factors such as obstacles, human movement, and ambient noise further complicate RSSI measurements (Guidara et al., 2018). Moreover, hardware variability, such as differences in microphone sensitivity and speaker output power, can introduce inconsistencies in measurements.

In summary, while RSSI-based acoustic positioning offers a low-cost and straightforward solution for indoor localization, its susceptibility to multipath effects, environmental variability, and hardware inconsistencies limit its effectiveness for precise localization. By integrating RSSI measurements with other techniques and employing advanced processing methods, the reliability and accuracy of indoor positioning systems can be significantly improved.

2.4 Conclusion

Acoustic positioning systems offer valuable solutions for indoor localization, overcoming limitations associated with conventional technologies like GPS, RFID, and UWB, particularly in scenarios involving complex indoor environments. This chapter reviewed fundamental acoustic propagation principles, highlighting critical phenomena such as attenuation, reflection, absorption, refraction, diffraction, and multipath propagation. A thorough understanding of these acoustic behaviors is essential, as they directly impact the accuracy and reliability of indoor positioning systems.

Time-based methods, including ToA and TDoA, offer precise localization capabilities but present synchronization challenges. ToA systems, despite their high potential accuracy, require rigorous synchronization between transmitters and receivers, making them sensitive to timing errors. In contrast, TDoA simplifies synchronization by relying solely on receiver-side synchronization, thus alleviating complexity on the transmitting side. However, both methods remain susceptible to multipath effects and NLoS propagation, necessitating sophisticated signal processing techniques such as pulse shaping and correlation-based filtering to mitigate errors caused by reflections and delayed paths.

AoA positioning provides directional information through microphone arrays without necessitating strict timing synchronization. This reduces system complexity, making AoA particularly advantageous for applications requiring directional guidance or coarse localization. Nonetheless, AoA systems demand precise microphone array calibration and are significantly affected by multipath propagation, where reflected signals can introduce substantial angular estimation errors. Additionally, the angular resolution and accuracy diminish with increasing distance from the anchor, limiting the practical deployment range unless larger arrays or hybrid solutions are used.

RSSI-based methods present a cost-effective, simple solution for localization through measuring signal amplitude. This approach is advantageous in scenarios requiring basic proximity detection or room-level localization. However, RSSI suffers heavily from multipath interference, environmental variability, and hardware inconsistencies, all of which introduce large localization uncertainties. Techniques such as averaging, fingerprinting, and hybrid systems integrating RSSI with ToA or AoA can enhance robustness and accuracy, yet precise localization remains challenging when relying solely on RSSI.

In summary, each acoustic positioning method discussed offers distinct strengths and faces unique challenges. Optimal indoor localization typically requires a hybrid approach that integrates complementary methods by combining the accuracy of ToA or TDoA, the directional capabilities of AoA, and the simplicity of RSSI. Such integrated systems, coupled with advanced signal processing and sensor fusion techniques, can significantly enhance the reliability, robustness, and precision of indoor acoustic positioning solutions, supporting diverse applications ranging from robotics and automation to smart-building technologies.

References

Filonenko, V., Cullen, C., & Carswell, J. (2013). Indoor positioning for smartphones using asynchronous ultrasound trilateration. *ISPRS International Journal of Geo-Information, 2*(3), 598–620. https://doi.org/10.3390/ijgi2030598

Guidara, A., Fersi G., Derbel, F., Ben Jemaa, M. (2018) Impacts of temperature and humidity variations on rssi in indoor wireless sensor networks procedia computer science. *126*, 1072–1081. https://doi.org/10.1016/j.procs.2018.08.044

Priyantha, N. B. (2005). *The cricket indoor location system.* Massachusetts Institute of Technology.

Yan, J., Zhao, W., Wu, Y. I., & Zhou, Y. (2023). Indoor sound source localization under reverberation by extracting the features of sample covariance. *Applied Acoustics, 210*, 109453. https://doi.org/10.1016/j.apacoust.2023.109453

Zhao, W., Goudar, A., & Schoellig, A. P. (2022). Finding the right place: Sensor placement for UWB time difference of arrival localization in cluttered indoor environments. *IEEE Robotics and Automation Letters, 7*(3), 6075–6082. https://doi.org/10.1109/LRA.2022.3165181

Chapter 3
System Architecture and Signal Processing

In this chapter, we discuss the overall system design for indoor acoustic positioning. We cover the physical architecture of sound-based positioning systems, the signal processing algorithms for distance measurement, the extension to three-dimensional localization, and techniques adopted from GPS systems to enable multiple devices to operate concurrently.

3.1 Design of Sound-Based Positioning Systems

Designing a sound-based positioning system involves careful consideration of the physical layout of devices, the roles of transmitters and receivers (i.e., system architecture), and the hardware components required (speakers, microphones, and synchronization mechanisms). This section discusses how the geometric arrangement of nodes and the choice of GPS-like vs. inverse GPS-like architectures affect system performance, as well as the key hardware elements and design trade-offs to implement a robust acoustic positioning system.

The geometric layout of transmitters (speakers or tweeters) and receivers (microphones) fundamentally impacts positioning performance. In a one-dimensional (1D) arrangement, devices are placed along a line (for example, down a hallway), which allows distance measurement along that axis but provides poor localization in other directions. A 2D configuration (e.g., anchors mounted on a ceiling or walls of a room) enables position estimation in a plane, suitable for tracking a device on a factory floor or in a greenhouse aisle. For full 3D coverage, transmitters and receivers are distributed in volumetric space (at different heights and locations) so that a target can be localized with depth (height) as well as horizontal position (Fig. 3.1b). Generally, achieving good geometry (similar to GPS satellite geometry) requires at least three non-collinear anchors for 2D, and four or more for 3D, to solve for the unknown coordinates. The arrangement should minimize geometric

Z. Huang, *Acoustic and Signal-Based Local Positioning*, Navigation: Science and Technology 18, https://doi.org/10.1007/978-981-95-6083-7_3

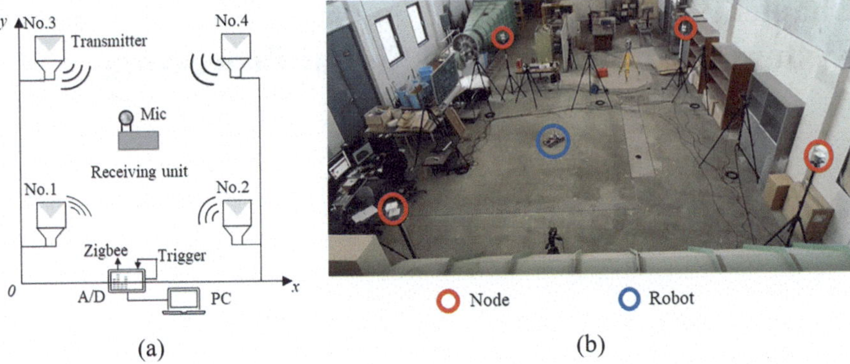

Fig. 3.1 A typical example of (**a**) the layout and (**b**) a physical setup of a sound-based indoor positioning system

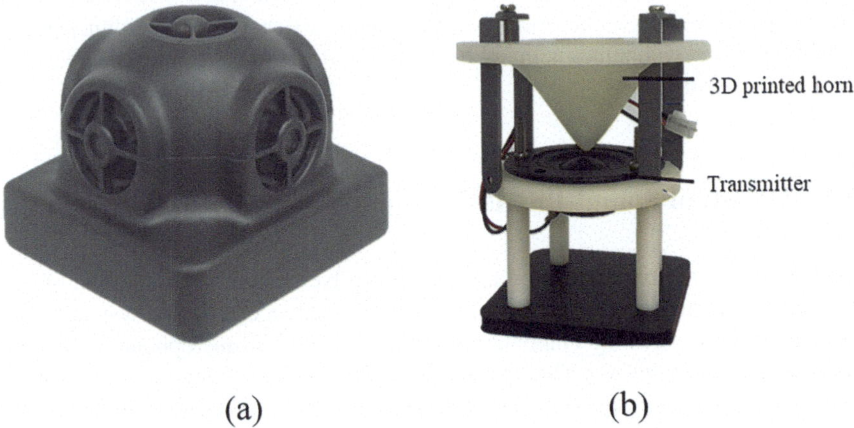

Fig. 3.2 Examples of transmitter design: (**a**) a circular array of speakers for 360° coverage, and (**b**) a single speaker outfitted with a 3D-printed omnidirectional horn to spread sound in all directions

dilution of precision by avoiding co-linear or co-planar placements that degrade accuracy.

Most ultrasonic transmitters exhibit a degree of directionality. In robotic applications where the orientation of the robot may change during operation, one effective approach is to employ a transmitter array (Fig. 3.2a) or a rotating speaker to achieve full 360° coverage. Another solution, illustrated in Fig. 3.2b, involves using a 3D-printed tapering structure that redirects the sound emitted from a single speaker uniformly in all directions, thereby creating an omnidirectional transmission pattern. In contrast, ceiling-mounted speakers are often equipped with directional emitters aimed downward to focus acoustic energy within a designated service area (e.g., the floor), minimizing undesired reflections from the ceiling. Ultimately, the choice between omnidirectional and directional emitters should be guided by the specific coverage requirements of the target environment.

An acoustic positioning system can be organized in one of two fundamental architectures: "GPS-like" (passive mobile) or "Inverse GPS-like" (active mobile) configurations (Fig. 3.3). The distinction lies in which devices transmit the acoustic signals and which devices listen.

GPS-like architecture has fixed transmitters and the mobile receiver (Fig. 3.3a). This configuration is analogous to GPS operation. Fixed-position speakers (anchors) are installed in the environment and periodically emit known localization signals (acoustic beacons). A microphone on the mobile agent (robot or tag) passively listens for these signals. The mobile node does not emit sound; it only receives. The arrival times of the signals from each anchor are used to compute the distances to those transmitters, and then the mobile can calculate its own position via trilateration. An advantage of the GPS-like approach is that it is easily extended to support multiple mobile agents simultaneously. Since anchor broadcasts can be heard by any number of receivers, many robots or tags can be localized concurrently by passively listening to the same set of beacons (making the system inherently scalable to multiple targets). Also, the mobile device can be kept lightweight and low-power (just a microphone and processing unit), while the power-hungry transmitters are fixed and can be connected to mains. However, a potential drawback is that signal strength may vary with the mobile's position. If the mobile is far from certain transmitters or located behind obstacles, it may experience attenuation or interference that affects measurement reliability. This architecture may also suffer from a near-far problem where a distant transmitter's signal might be much weaker compared to a nearer one.

Inverse GPS-like architecture has the mobile transmitter and fixed receivers (Fig. 3.3b). In this configuration, the roles are reversed. The mobile platform (e.g., a robot or drone) carries an active speaker that emits the acoustic signal, and an array of microphones is placed at fixed anchor locations around the environment's

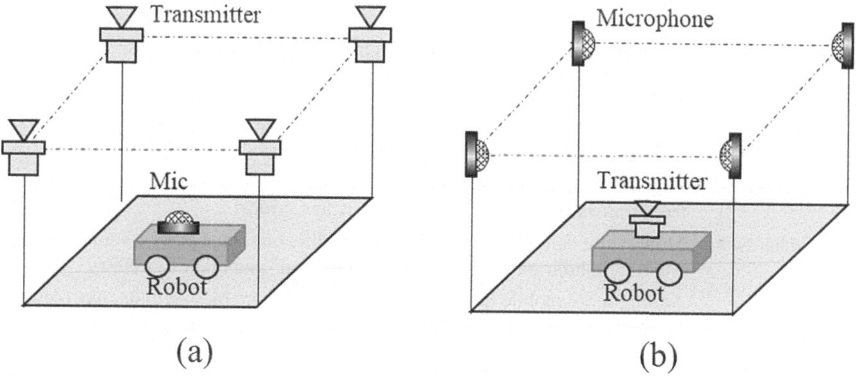

Fig. 3.3 Configurations of the acoustic positioning system: (**a**) GPS-like architecture, and (**b**) inverse GPS-like architecture

perimeter to receive the signal. Each anchor detects the arrival time of the sound pulse, and these timestamps are sent to a central processor (or the robot) which computes the distances and the mobile's position. One key advantage of the inverse approach is that it mitigates the near-far problem presented in the GPS-like setup. Because the mobile transmitter provides a strong signal wherever it goes, all anchors receive signals of comparatively similar strength regardless of the mobile's location in the space. This can ensure more consistent detection reliability in enclosed environments such as greenhouses (Widodo et al., 2013). Another benefit is that anchors (microphone units) can be relatively simple and low-power, potentially just recording devices sending data to a central unit, while the mobile does the active work. The downside lies in scalability, since an inverse GPS-like system can typically support only one, or at most a few, active mobile transmitters in the area at the same time. If multiple mobiles tried to emit simultaneously, their signals could collide or interfere. To handle multiple mobiles, the system would need to coordinate time slots or use distinct signal codes/frequencies for each, which adds complexity. Thus, this architecture is less scalable when many targets need to be tracked concurrently. On the other hand, for a single robot in a known area, it offers improved consistency and potentially better accuracy in a controlled environment (Table 3.1).

In summary, each architecture comes with trade-offs. The GPS-like (anchor-to-mobile) system scales well to many receivers/tags and keeps mobiles simple, but may suffer if the mobile moves far from some transmitters or if the environment causes interference. The inverse GPS-like (mobile-to-anchors) system provides robust signal coverage for the one active node and uniform signal strength at anchors, but it is not as naturally equipped to handle many simultaneous transmitters and requires the mobile to carry the power-consuming emitter. Designers must weigh these trade-offs when choosing the system structure. In practice, the positioning infrastructure often includes a central controller or server that coordinates the measurements. For example, in a GPS-like system the controller may schedule or trigger the anchor transmissions or broadcast a synchronization signal, while in an inverse system it may collect timing data from all the receiving anchors. Either way,

Table 3.1 Comparison between GPS-like and inverse GPS-like configurations

Feature	GPS-like	Inverse GPS-like
Configuration	Fixed transmitters and the mobile receiver	Mobile transmitter and fixed receivers
Synchronization	Requires synchronization among fixed transmitters	Synchronization among fixed receivers is less critical
Mobile hardware	Mobile unit can be very simple (e.g. one microphone and a processor)	Mobile unit must carry a sound emitter (speaker) and related circuitry
Scalability	Adding more mobile units is straightforward	Adding more mobile units may require coordination to prevent signal interference
Use case suitability	Suitable for scenarios with multiple mobile units and fixed infrastructure	Suitable for scenarios with a single mobile unit and multiple fixed receivers

time synchronization (discussed later in Sect. 3.4) is critical for comparing arrival times.

3.2 Signal Detection and Processing Algorithms for Distance Measurement

ToA (Time-of-Arrival) methods are widely adopted in sound-based positioning systems due to their high accuracy and cost-effectiveness. In this section and the next, we examine the development of such systems, with core implementation functions for a ToA-based indoor positioning system provided in the Appendix. This section focuses on how distances are measured via acoustic signals, and Sect. 3.3 will extend this to full 3D position estimation.

The basic process of the ToF method begins with the speaker emitting two signals simultaneously. One is a high-speed trigger signal that uses RF waves, and the other is a spread spectrum sound signal referred to as spread spectrum sound (SSSound). The RF trigger signal, due to its significantly higher propagation speed, is detected almost instantaneously by the microphone, which then initiates the recording of the incoming SSSound signal. The time difference between the reception of the RF trigger and the arrival of the SSSound signal constitutes the ToF. Multiplying this time interval by the speed of sound, adjusted for environmental conditions, yields the distance between the speaker and microphone.

The SSSound signal is meticulously designed to facilitate accurate cross-correlation and ToF estimation. Its design is based on a Maximum Length Sequence (M-sequence), a type of pseudorandom binary sequence renowned for its exceptional autocorrelation properties. M-sequences exhibit a sharp peak at zero lag in their autocorrelation function, with minimal values elsewhere, facilitating precise time delay measurements. While other pseudorandom sequences like Kasami and Gold sequences are also considered in signal processing applications, M-sequences are particularly advantageous for ToF-based distance measurements due to their superior autocorrelation characteristics. Kasami sequences offer good cross-correlation properties, making them suitable for systems requiring multiple access capabilities. Gold sequences, generated by the modulo-2 addition of two M-sequences, provide a balance between autocorrelation and cross-correlation properties, which is beneficial in certain communication systems. However, for single-source distance estimation tasks, the distinct autocorrelation peak of M-sequences ensures more accurate and reliable ToF calculations (Sawlikar & Sharma, 2011).

To enhance the signal's robustness and facilitate transmission through the acoustic medium, the M-sequence is modulated using Binary Phase Shift Keying (BPSK) onto a carrier wave. The generation of this SSSound is illustrated in Fig. 3.4. The resulting SSSound signal is defined by the following equations:

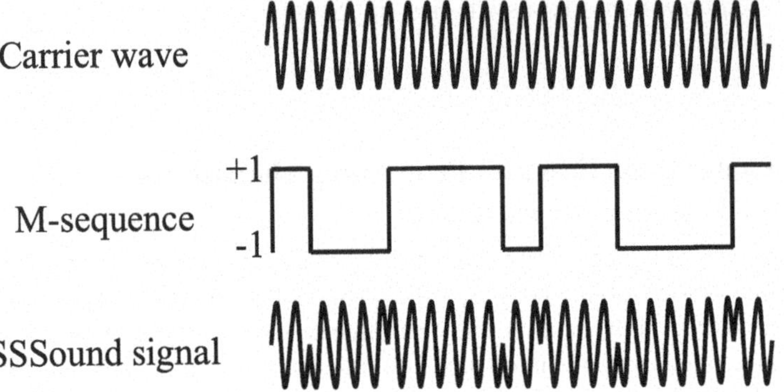

Fig. 3.4 Generation of the spread-spectrum sound (SSSound) by modulating an M-sequence with a carrier wave (BPSK modulation)

Table 3.2 Typical parameters for an SSSound signal (example)

Property	Value/remark
Sampling frequency (f_s)	96 kHz
Sampling bit	16
M-sequence length	1024
Modulation	BPSK
Carrier wave frequency (f_c)	24 kHz
Chip rate (f_{chip})	12 kcps

$$s(n) = \text{round}\left\{\sin\left[\frac{2\pi f_c (n-1)}{f_s}\right] \times M(ms)\right\} \tag{3.1}$$

with

$$\text{ms} = \text{floor}\left[\frac{f_{chip}(n-1)}{f_s}\right] \tag{3.2}$$

where, $s(n)$ denotes the discrete SSSound signal sample, f_c is the carrier frequency, f_s is the sampling frequency, f_{chip} is the chip rate, $M(ms)$ represents the value of the M-sequence at index ms, and n is sample index.

Table 3.2 summarizes the characteristics of a typical SSSound signal for the inverse GPS-like configuration (Huang et al., 2017). This signal spans a broad frequency range from 12 to 36 kHz, enabling effective ToF measurements. The transmitter is mounted on a mobile robot, facilitating precise location tracking. This configuration achieves an accuracy of approximately 2 cm within a 30 × 30 m area (Widodo et al., 2014). Moreover, by operating above the typical human hearing range, the system minimizes acoustic disturbances in occupied environments.

The frequency spectrum of the typical SSSound signal spans a range that includes both the carrier frequency and its side lobes (Fig. 3.5). The main lobe is between 12

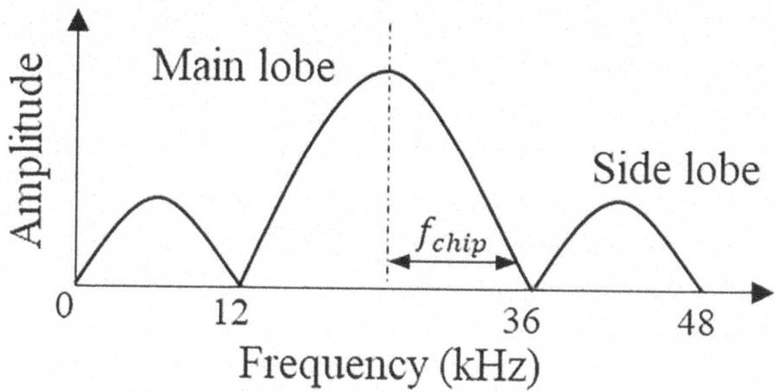

Fig. 3.5 Frequency spectrum of the SSSound signal (schematic)

and 36 kHz. Side lobes below 12 kHz and above 36 kHz are filtered out to reduce noise impact. However, environmental noise, particularly in the audible range (0–12 kHz), can interfere with signal detection. To mitigate this, the side lobes within 0–12 kHz and 36–48 kHz are attenuated during cross-correlation calculations. This filtering preserves the main lobe of the signal, maintaining its autocorrelation properties while reducing susceptibility to noise.

Accurate estimation of the ToF requires precise determination of the arrival time of the SSSound signal. This is achieved through cross-correlation between the transmitted SSSound signal and the received signal at the microphone. The cross-correlation function $c(t)$ is computed as:

$$c(t) = \sum_{n=0}^{N-1} s(n) r(n+t) \tag{3.3}$$

where, $s(n)$ denotes the transmitted SSSound signal, $r(n)$ denotes the received signal, N is the length of the signal, and t is the time lag.

The peak of the cross-correlation function corresponds to the point of maximum similarity between the transmitted and received signals, indicating the arrival time of the SSSound signal.

To distinguish the true peak from potential noise-induced peaks, a threshold c_{th} is established:

$$c_{th} = C_{ave} + 4\sigma_{corr} \tag{3.4}$$

where, C_{ave} is the average absolute value of the cross-correlation, and σ_{corr} is the standard deviation of the cross-correlation.

Direct computation of cross-correlation in the time domain can be computationally intensive, especially for long signals. To enhance efficiency, the cross-correlation is computed in the frequency domain using the Fast Fourier Transform (FFT) and its inverse (IFFT).

Steps for FFT-Based Cross-Correlation:

(a) **Zero-Padding**: To prevent circular convolution effects and ensure linear cross-correlation, both the transmitted signal $s(n)$ and the received signal $r(n)$ are zero-padded. The length of the zero-padded signals should be at least $N = length(s) + length(r) - 1$.

(b) **Compute FFTs**:

Calculate the FFT of the zero-padded transmitted signal: $S(f) = FFT(s(n))$.
Calculate the FFT of the zero-padded received signal: $R(f) = FFT(r(n))$.

(c) **Compute Cross-Spectrum**:

Multiply $S(f)$ with the complex conjugate of $R(f)$: $C(f) = S(f) \cdot R^*(f)$.

(d) **Compute Inverse FFT**:

Apply the inverse FFT to $C(f)$ to obtain the cross-correlation sequence: $c(t) = IFFT(C(f))$.

The peak of the cross-correlation sequence $c(t)$ indicates the time delay Δt between the transmitted and received signals. This time delay corresponds to the ToF of the SSSound signal.

Figure 3.6 illustrates the cross-correlation between the transmitted and received signals, clearly displaying a distinct peak at C_{max}. This peak indicates the point of maximum similarity and corresponds to the arrival time of the SSSound signal. The associated sample index C_{max} represents the ToF, which can be converted to actual time using the sampling interval (for instance, at 96 kHz sampling, 1 sample ≈ 10.4 μs).

Once the time delay Δt is determined, the distance d_i between the speaker and microphone can be calculated using the speed of sound v_s:

$$v_s = 331.5 + 0.61\left(\frac{T_i + T_s}{2}\right) \tag{3.5}$$

$$d_i = v_s \Delta t - t_{delay} \tag{3.6}$$

where, T_i is the temperature at the microphone, T_s is the temperature at the speaker, v_s is the speed of sound in air (m/s), Δt is the time delay (s), d_i is the distance between speaker i and microphone (m), and t_{delay} is the time delay offset caused by the High Pass Filter (HPF) circuit of the speaker (s).

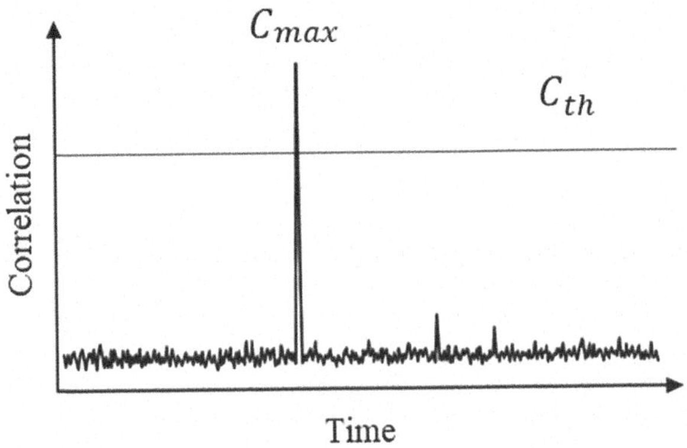

Fig. 3.6 Cross-correlation between the transmitted and received signals

This FFT-based approach significantly reduces computational complexity from N^2 to $NlogN$, enabling real-time processing and efficient distance estimation in the sound-based positioning system.

3.3 Integration of Positioning Methods for 3D Estimation

In sound-based positioning systems, accurately determining the 3D location of a receiver (e.g., a microphone) is crucial for various applications, including indoor navigation and robotics. The ToF measurements, which calculate the time taken by a sound signal to travel from a transmitter (speaker) to a receiver, serve as the foundational data for such positioning. By obtaining multiple ToF measurements from different known transmitter locations, it's possible to estimate the receiver's 3D position through trilateration techniques.

Trilateration is a geometric method that determines the position of a point based on its distances from three or more known locations. In a 3D space, at least three distance measurements are typically required to resolve the receiver's position accurately. Each distance measurement defines a sphere centered at the known transmitter location, with a radius equal to the measured distance. The intersection point of these spheres corresponds to the receiver's position. Figure 3.7 illustrates this concept. Each transmitter defines a sphere (with radius equal to the measured distance to the receiver). The black point indicates the estimated receiver position at the intersection of these spheres. The receiver's coordinates are found at the intersection of multiple spheres.

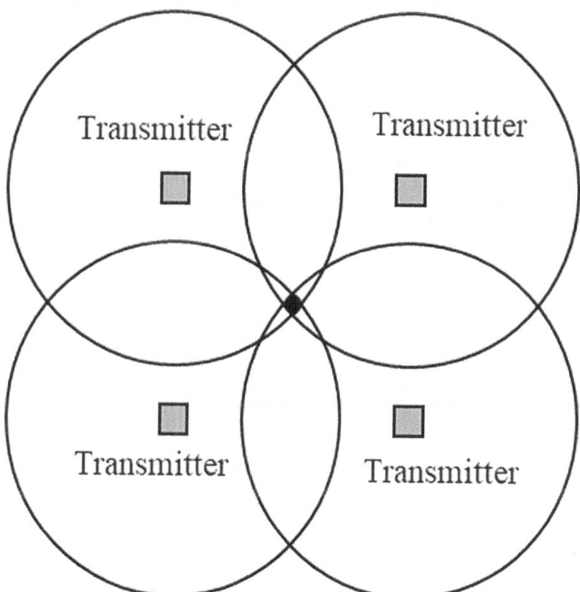

Fig. 3.7 Position estimation by trilateration in 3D space

Let:
$S = (x_m, y_m, z_m)$ be the unknown position of the receiver.
$M_i = (x_i, y_i, z_i)$ be the known positions of the transmitters, where $i = 1, 2, 3, 4$.
d_{si} be the measured distance from transmitter i to the receiver.
The Euclidean distance between the receiver and each transmitter is given by:

$$d_{si} = \sqrt{(x_m - x_i)^2 + (y_m - y_i)^2 + (z_m - z_i)^2} \tag{3.7}$$

These equations are nonlinear due to the square root and squared terms. To solve for the receiver's position, we employ an iterative least squares method, which linearizes the equations and refines the position estimate through successive approximations.

The iterative least squares approach begins with an initial guess of the receiver's position, $S^{(0)} = \left(x_m^{(0)}, y_m^{(0)}, z_m^{(0)}\right)$. Using this estimate, we calculate the estimated distances to each transmitter:

$$d_{si}^{(0)} = \sqrt{\left(x_m^{(0)} - x_i\right)^2 + \left(y_m^{(0)} - y_i\right)^2 + \left(z_m^{(0)} - z_i\right)^2} \tag{3.8}$$

The difference between the measured and estimated distances is:

$$\Delta d_{si} = d_{si} - d_{si}^{(0)} \tag{3.9}$$

These differences are used to compute corrections to the initial position estimate. We linearize the distance equations using a first-order Taylor series expansion:

$$\Delta d_{si} \approx \frac{\partial d_{si}^{(0)}}{\partial x_m} \Delta x_m + \frac{\partial d_{si}^{(0)}}{\partial y_m} \Delta y_m + \frac{\partial d_{si}^{(0)}}{\partial z_m} \Delta z_m \tag{3.10}$$

where the partial derivatives are:

$$\frac{\partial d_{si}^{(0)}}{\partial x_m} = \frac{x_m^{(0)} - x_i}{d_{si}^{0}}$$

$$\frac{\partial d_{si}^{(0)}}{\partial y_m} = \frac{y_m^{(0)} - y_i}{d_{si}^{0}}$$

$$\frac{\partial d_{si}^{(0)}}{\partial z_m} = \frac{z_m^{(0)} - z_i}{d_{si}^{0}}$$

By assembling these equations for all transmitters, we form a system of linear equations:

$$\Delta d = A \cdot \Delta S \tag{3.11}$$

where, $A = \begin{bmatrix} \frac{x_m^{(0)} - x_1}{d_{s1}^{(0)}} & \frac{y_m^{(0)} - y_1}{d_{s1}^{(0)}} & \frac{z_m^{(0)} - z_1}{d_{s1}^{(0)}} \\ \frac{x_m^{(0)} - x_2}{d_{s2}^{(0)}} & \frac{y_m^{(0)} - y_2}{d_{s2}^{(0)}} & \frac{z_m^{(0)} - z_2}{d_{s2}^{(0)}} \\ \frac{x_m^{(0)} - x_3}{d_{s3}^{(0)}} & \frac{y_m^{(0)} - y_3}{d_{s3}^{(0)}} & \frac{z_m^{(0)} - z_3}{d_{s3}^{(0)}} \\ \frac{x_m^{(0)} - x_4}{d_{s4}^{(0)}} & \frac{y_m^{(0)} - y_4}{d_{s4}^{(0)}} & \frac{z_m^{(0)} - z_4}{d_{s4}^{(0)}} \end{bmatrix}$.

Here, Δd is the vector of distance differences. A is the matrix of partial derivatives (Jacobian matrix). $\Delta S = (\Delta x_m, \Delta y_m, \Delta z_m)^T$ is the correction vector for the receiver's position.

To solve for ΔS, we apply the least squares solution:

$$\Delta S = \left(A^T A\right)^{-1} A^T \Delta d \tag{3.12}$$

The receiver's position estimate is then updated:

$$S^{(1)} = S^{(0)} + \Delta S \tag{3.13}$$

This iterative process continues until the corrections ΔS are below a predefined threshold, indicating convergence. Typically, 50 iterations are sufficient to achieve an accurate position estimate.

3.4 Features of GPS-Like System: TDMA and FDMA Approaches

In a GPS-like indoor acoustic positioning system, multiple fixed transmitters emit acoustic signals, and mobile devices equipped with microphones measure the received signals' ToF to determine their positions. Unlike the inverse GPS-like system, the GPS-like configuration involves multiple transmitters emitting signals simultaneously or sequentially, potentially causing signal interference. This interference is described as near-far problem. In Fig. 3.8, the stronger signal from a nearby transmitter can mask the weaker signal from a far transmitter at the receiver, if both arrive at the same time or frequency band. To mitigate this interference and expand the system's capability for multiple target positioning, two signal multiplexing strategies are employed: Time-division Multiple Access (TDMA) and Frequency-division Multiple Access (FDMA).

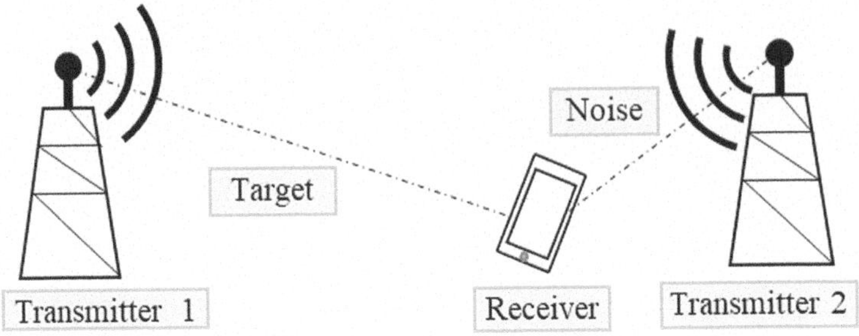

Fig. 3.8 Illustration of the near-far problem in a GPS-like acoustic system

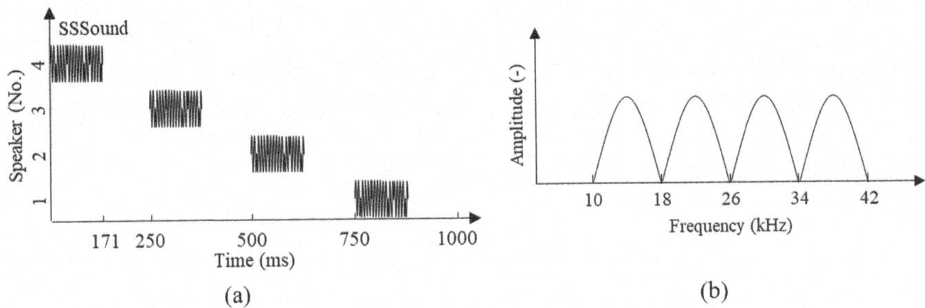

Fig. 3.9 Illustration of multiplexing approaches for multiple transmitters: (**a**) TDMA—each transmitter emits in a designated time slot (no overlap in time); (**b**) FDMA—all transmitters emit together but on different carrier frequencies (separated in frequency domain)

In a TDMA scheme, each transmitter is assigned a distinct time slot during which it is allowed to transmit its acoustic signal. By staggering the transmissions in time, overlaps are avoided. For example, the system employs TDMA slots based on the SSSound signals described earlier. With four transmitters, a 2047-length M-sequence is assigned to each, using a 24 kHz carrier and a 12 kcps chip rate. This configuration yields an individual signal length of approximately 171 ms (Huang et al., 2020). Figure 3.9a illustrates the timing of TDMA signals: each transmitter emits in sequence, one after the other. In our implementation, we allocate a 0.25 s interval between the start of one transmission and the start of the next, so that by the time the second transmitter begins, the first transmitter's 171 ms signal has finished propagating in the environment (with some guard time to spare). With four transmitters, this results in a total TDMA frame of approximately 1 s (0.25 s × 4). This carefully timed approach prevents overlap and ensures that a receiver always processes one signal at a time, yielding clear, distinct ToF measurements for each anchor. However, TDMA introduces a trade-off in that the update rate for each transmitter's distance measurement is lower. Each anchor transmits only once per frame, which results in an update rate of 1 Hz if the frame duration is 1 s.

In an FDMA scheme, all transmitters can emit simultaneously, but each uses a different carrier frequency band. The acoustic signals are thus separated in frequency rather than in time. In our example, we use four distinct carrier frequencies for four transmitters: 17 kHz, 26 kHz, 35 kHz, and 44 kHz, respectively (spaced out to minimize overlap). Each transmitter still uses a pseudorandom M-sequence modulation, but because the bandwidth per channel is limited, the parameters are adjusted. In FDMA, a shorter sequence of length 1023 and a lower chip rate of 4 kcps are used for each channel, which stretches each signal to a duration of about 256 ms. Figure 3.9b shows a conceptual timing diagram for FDMA, where all four signals occur concurrently but on separate frequency channels. A receiver (with a broadband microphone) records the composite signal, and then applies band-pass filtering or frequency-domain processing to isolate each channel. The advantage of FDMA is that all distances can be measured simultaneously (in principle yielding a faster update rate, since every time frame contains all signals). Additionally, the longer signal length (256 ms) and lower chip rate in our FDMA example can improve resilience to noise (more integration time per bit). The challenge, however, is that the receiver's signal processing is more complex and the channels are not completely independent: some cross-talk can occur if the frequency responses overlap or if nonlinearities introduce intermodulation. Still, with careful channel separation, FDMA allows concurrent multi-anchor operation without time multiplexing.

An important consideration in these multi-transmitter systems is how the signal processing (especially correlation-based ToF detection) is affected by the multiplexing scheme. Accurate ToF detection relies on a sharp and distinct correlation peak for each transmitted signal at the receiver. In practice, we observe differences between TDMA and FDMA in terms of the correlation peak shape and detectability. The length of the SSSound signal (L_S, unit s) can be computed using the relationship, which is determined by the length of M-sequence (M_{length}) and chip rate value (f_{chip}):

$$L_S C(t) = \sum_{n=0}^{N-1} s(n) r(n+t) = \frac{M_{length}}{f_{chip}} \tag{3.14}$$

A lower chip rate and a longer M-sequence length improve noise resilience, although they simultaneously result in longer signal duration.

In typical SSSound signals with wide frequency spectrum range, the cross-correlation function exhibits a sharp and narrow peak. For instance, with a chip rate of 12 kcps and a 2047-length M-sequence, the autocorrelation peak spans 16 samples (Fig. 3.10a). Additionally, due to the carrier wave effect, three distinct peaks are observed within this width, spaced four samples apart, corresponding to the carrier wavelength. The central peak accurately indicates the arrival time of the SSSound signal, facilitating precise ToF measurements.

Conversely, FDMA signals, often use lower chip rates to accommodate longer signal durations. This results in broader cross-correlation peaks with multiple subpeaks exceeding the detection threshold (Fig. 3.10b). The widened and more complex peak structure complicates the identification of the exact arrival time, especially in noisy environments where ambient noise can interfere with the correlation peak's

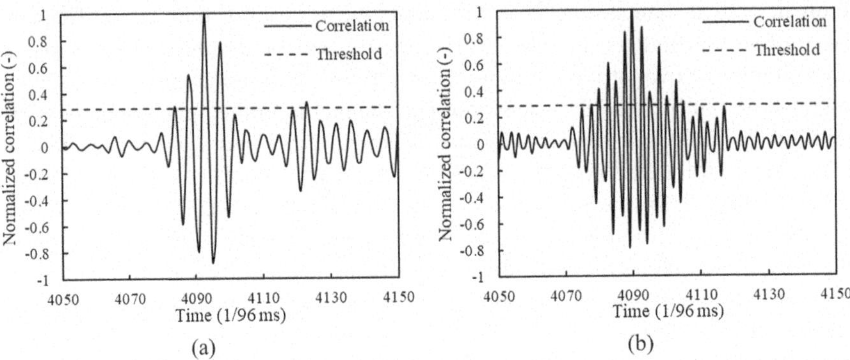

Fig. 3.10 Examples of cross-correlation results: (**a**) for a wideband TDMA signal (high chip rate) showing a sharp, well-defined peak; (**b**) for an FDMA signal (lower chip rate per channel) showing a broader peak with multiple sub-peaks above the threshold

shape. The broader peaks in FDMA systems are attributed to the lower chip rates and the inherent properties of the carrier frequencies used. Lower chip rates increase the signal's susceptibility to noise, and the varying phase shifts introduced by different carrier frequencies can distort the correlation function. These factors collectively make it challenging to achieve the same level of ToF detection accuracy as in systems utilizing SSSound signals with higher chip rates.

Therefore, system designers must carefully balance these parameters to optimize the trade-off between noise tolerance, multi-target capability, and time resolution. TDMA offers simplicity and sharp correlation characteristics at the cost of update rate, whereas FDMA offers simultaneous operation but demands more sophisticated signal processing. In practice, hybrid approaches or code-division techniques can also be considered, but those require careful code design to maintain low cross-correlation between signals. In summary, depending on the specific application requirements (number of devices to track, required update frequency, environmental noise levels, etc.), one should choose the appropriate multiplexing strategy. The architectures and processing techniques described in this chapter provide a foundation that can be adapted or combined to achieve robust indoor acoustic positioning in a variety of scenarios.

References

Huang, Z., Ono, M., Shiigi, T., Suzuki, T., Harshana, H., Nakanishi, H., & Kondo, N. (2017). Is spread spectrum sound a robust local positioning system for a quadcopter operating in a greenhouse? *Chemical Engineering Transactions, 58*, 829–834. https://doi.org/10.3303/CET1758139

Huang, Z., Tsay, L. W. J., Shiigi, T., Zhao, X., Nakanishi, H., Suzuki, T., Ogawa, Y., & Kondo, N. (2020). A noise tolerant spread Spectrum sound-based local positioning system for operating a quadcopter in a greenhouse. *Sensors (Basel), 20*(7), 7. https://doi.org/10.3390/s20071981

Sawlikar, A., & Sharma, M. (2011). Analysis of different pseudo noise sequences. *International Journal of Computer Technology and Electronics Engineering, 1*(2), 156–161.

Widodo, S., Shiigi, T., Than, N. M., Kurosaki, H., Kikuchi, H., Yanagida, K., Nakatsuchi, Y., Ogawa, Y., & Kondo, N. (2013). Design of Self-calibration Method for sound-based positioning system in greenhouse. *IFAC Proceedings Volumes, 46*(4), 4. https://doi.org/10.3182/20130327-3-JP-3017.00075

Widodo, S., Shiigi, T., Than, N. M., Kikuchi, H., Yanagida, K., Nakatsuchi, Y., Ogawa, Y., & Kondo, N. (2014). Wind compensation for an open field spread spectrum sound-based positioning system using a base station configuration. *Engineering in Agriculture, Environment and Food, 7*(3), 3. https://doi.org/10.1016/j.eaef.2014.04.001

Chapter 4
Doppler Effect and Compensation in Acoustic Positioning Systems

Spread spectrum sound-based local positioning systems (SSSLPS) have demonstrated centimeter-level accuracy in static scenarios and even for moving ground robots. However, when either the transmitter or the receiver is in motion, a major challenge arises because the Doppler effect induces a frequency shift in the acoustic signals, which can significantly degrade ranging accuracy. Similar issues occur for UAV-based SSSLPS implementations, where Doppler-induced ToA errors must be mitigated to maintain the required accuracy.

This chapter focuses on understanding the Doppler effect in sound-based positioning and on modern algorithms to compensate for Doppler shifts. Section 4.1 begins with an explanation of the Doppler effect in the context of acoustic positioning for moving robots and UAVs, including relevant equations and an illustrative figure demonstrating how motion shifts the frequency of sound waves. In Sect. 4.2, we present three state-of-the-art Doppler shift compensation (DSC) algorithms that enable accurate acoustic localization of moving platforms. The first algorithm (Sect. 4.2.1) uses a pilot tone–based approach to estimate and compensate Doppler shifts by tracking the frequency offset of a known reference signal embedded in the acoustic transmission. The second algorithm (Sect. 4.2.2) uses a carrier-frequency extraction technique based on a digital square-law detector to measure and correct Doppler shifts in an SSSLPS employing frequency-division multiple access (FDMA). The third algorithm (Sect. 4.2.3) takes a different approach: it estimates the relative velocity and searches across a range of frequency-shifted reference signals, selecting the one that yields the strongest correlation (and thus the most accurate ToA). All algorithms are described in detail with their methodology, equations, and processing flowcharts, followed by summaries of experimental results demonstrating their efficacy. Finally, Section 4.3 provides a brief conclusion highlighting the significance of these Doppler compensation techniques for high-accuracy acoustic positioning under motion, and discusses remaining challenges for future research.

Z. Huang, *Acoustic and Signal-Based Local Positioning*, Navigation: Science and Technology 18, https://doi.org/10.1007/978-981-95-6083-7_4

4.1 Doppler Effect in Moving Platforms

When either the source or the receiver of a sound signal is moving, the observed frequency of the signal is shifted relative to the emitted frequency. This phenomenon, known as the Doppler effect, is critical in acoustic positioning because most localization systems assume a known signal frequency for accurate time-of-flight measurements (Fig. 4.1). In essence, if a robot carrying a microphone moves toward a fixed sound beacon, it will encounter the sound wave's crests more frequently than if it were stationary—resulting in an upward shift in the observed frequency. Conversely, if the robot is moving away, the observed frequency is lower. (A classic example is the pitch of a siren sounding higher as it approaches and lower as it recedes.) The Doppler frequency shift for a moving observer (receiver) and stationary source can be quantified by:

$$f_{obs} = \left(\frac{v \pm v_r}{v} \right) f_{src} \tag{4.1}$$

where, f_{obs} is the observed frequency at the receiver, f_{src} is the emitted (source) frequency, v is the speed of sound in air (typically ~343 m/s), and v_r is the velocity of the receiver relative to the source.

Equation (4.1) shows that a receiver moving toward the source ($+v_r$) observes a higher frequency, while moving away ($-v_r$) yields a lower frequency. For a moving source and stationary receiver, a similar relationship holds: $f_{obs} = f_{src}(v/(v \mp v_s))$, where v_s is the source velocity (the sign convention is such that a source moving toward the receiver increases f_{obs}. In both cases, the frequency shift $\Delta f = f_{obs} - f_{src}$ is approximately $\pm(v_r/v)\, f_{src}$ for $v_r \ll v$. For instance, a robot moving at 0.5 m/s ($\approx$1.8 km/h) toward a 24 kHz ultrasonic beacon will observe roughly a +35 Hz frequency shift (since $v_r/v \approx 0.0015$), whereas moving at 0.5 m/s away would cause a −35 Hz shift. Though this fractional change (~0.15%) in frequency may seem small, it has significant impacts on spread-spectrum acoustic ranging.

In SSSLPS (as discussed in the previous chapter), signals are often encoded with pseudorandom sequences (e.g., M-sequences) and emitted at ultrasonic carrier

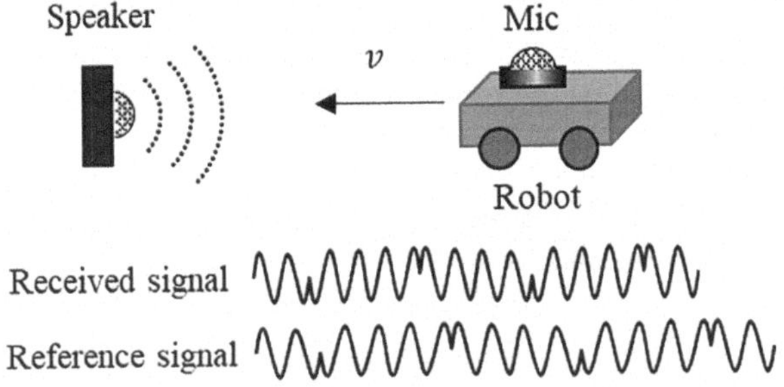

Fig. 4.1 Illustration of the Doppler effect for a moving source and stationary observer

frequencies (typically 20–40 kHz) to allow precise correlation-based timing. Doppler shifts distort both the carrier and the effective chip rate of the spreading sequence, causing the received signal to lose alignment with the expected reference signal. The result is a weaker correlation peak and a biased or ambiguous ToA measurement if not corrected. A correlation signal-to-noise ratio (SNR_{corr}) to quantify how strong and clear the peak is. A typical definition used is:

$$SNRcorr = 20\log\frac{C_p}{C_m} \tag{4.2}$$

where C_p is the amplitude of the correlation's peak, and C_m is a measure of the background correlation level (for example, the average absolute correlation or the standard deviation of the correlation values).

An instance from the fourth dynamic sample (Fig. 4.2a) recorded during movement at 400 mm/s shows that the corresponding signal-to-noise ratio of the correlation peak (*SNRcorr*) values without and with compensation were 6.8 and 58.6, respectively. Following application of the DSC algorithm, the correct arrival time of the sound signal was successfully identified (Fig. 4.2b). This demonstrates that Doppler compensation is crucial for achieving reliable time-of-arrival (ToA) estimation.

This effect grows worse with increasing relative velocity. Empirical studies have shown that as the relative speed between transmitter and receiver increases, the *SNRcorr* steadily declines. One study reported that beyond a relative speed of approximately 0.20 m/s, the correlation peak could no longer be reliably detected (Fig. 4.3) (Huang et al., 2021). In other words, without Doppler compensation, a moving robot or UAV may drop out of the acoustic tracking system once its speed exceeds a certain threshold. Even at lower speeds, uncompensated Doppler shifts introduce ranging errors roughly proportional to the frequency offset. Clearly,

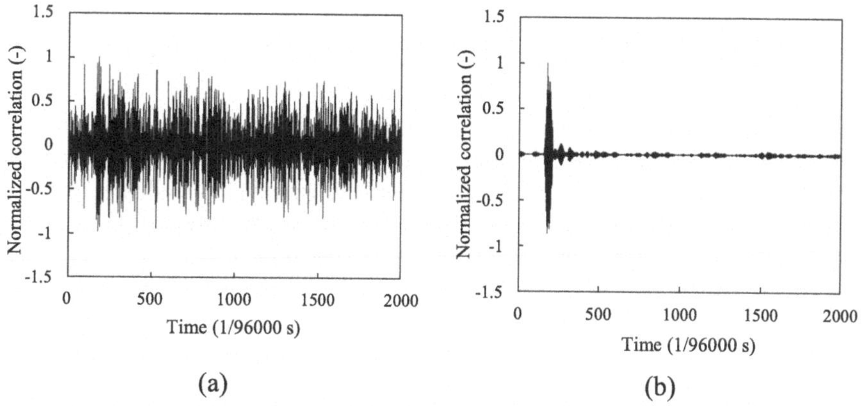

Fig. 4.2 Example of correlation (**a**) without and (**b**) with the DSC

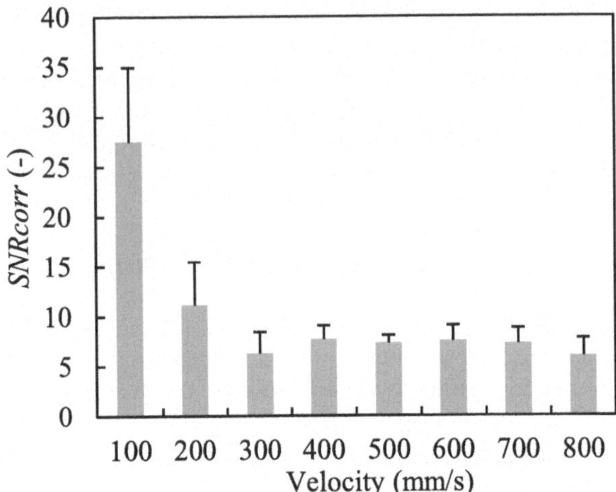

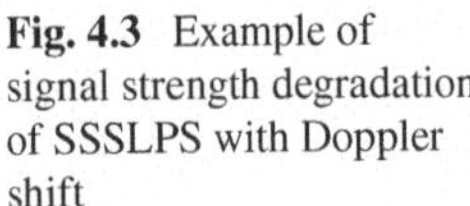
Fig. 4.3 Example of signal strength degradation of SSSLPS with Doppler shift

compensating for Doppler shifts is essential for maintaining high accuracy in dynamic acoustic positioning.

Researchers have explored several strategies to handle Doppler effects in sound-based systems. One approach is to apply signal-processing filters (for example, extended Kalman filters (Ko et al., 2008) or particle filters (Smith et al., 2004)) to track and compensate for Doppler as part of the distance estimation process. These filter-based methods can smooth gradual Doppler variations, but their accuracy in fast dynamic scenarios has generally been limited to about 0.2 m of error, which is insufficient for applications that require precision on the order of only a few centimeters.

In light of these limitations, recent research has shifted toward more direct and adaptive Doppler compensation techniques embedded within the SSSLPS processing pipeline. The following sections introduce three such techniques in detail.

4.2 Doppler Shift Compensation Algorithms

For accurate localization of moving robots using sound, Doppler-induced frequency shifts must be compensated when measuring distances. In this section, we present three Doppler Shift Compensation techniques recently proposed to address motion-induced errors in SSSLPS. We begin with a brief overview of related prior work, and then detail each algorithm's methodology and performance.

4.2.1 *Doppler Shift Compensation Using Pilot Tone Method (Algorithm 1)*

The pilot tone method offers an efficient and robust approach to estimating and compensating for Doppler shifts in acoustic positioning systems (Widodo et al., 2013). Unlike blind estimation methods, it relies on embedding a known reference tone into the transmitted signal. This known pilot tone enables precise measurement and correction of Doppler-induced frequency deviations directly at the receiver.

A practical implementation of the pilot-tone method involves several clearly defined signal-processing stages (Fig. 4.4):

(a) **Pilot Tone Transmission**

A carefully chosen single-frequency pilot tone is generated and summed with the primary transmitted data signal. A pilot tone is typically a continuous, single-frequency sinusoidal wave included alongside the primary ranging signal. Its frequency, f_p, is precisely known and carefully selected to avoid interference with data-carrying portions of the signal. Usually, the pilot tone is placed at the edges or within guard bands of the frequency spectrum. In practical implementations, the pilot's amplitude is carefully managed to minimize its impact on overall signal power and bandwidth utilization. The transmitted SSSound wave $s(n)$ is generated in the following equation:

$$s(n) = \text{round}\left\{\sin\left[\frac{2\pi f_c(n-1)}{f_s}\right] \times M(ms)\right\} + \sin\left[\frac{2\pi f_p(n-1)}{f_s}\right] \tag{4.3}$$

(b) **Reception and Filtering**

At the receiver, extracting the pilot tone accurately is essential. Initially, the incoming signal undergoes synchronization procedures, which often rely on distinct synchronization symbols or sequences separate from the pilot tone. Once coarse synchronization is achieved, a bandpass filter (BPF) isolates the pilot tone from the rest of the received spectrum. The BPF's bandwidth accommodates the maximum anticipated Doppler shift. For example, a typical setup for a pilot at 36 kHz might use a BPF of approximately 500 Hz bandwidth to

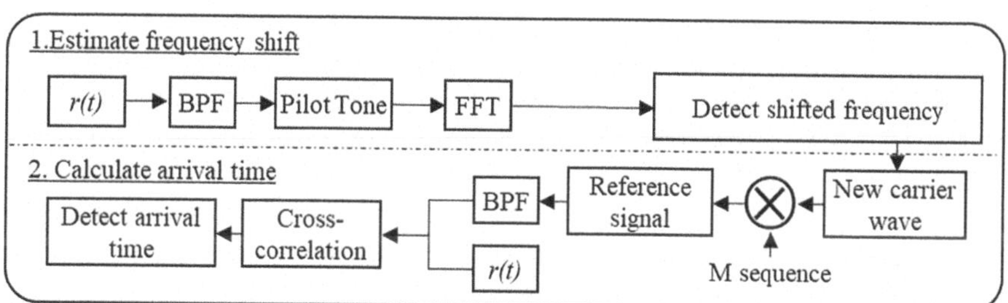

Fig. 4.4 Block diagram of the Doppler Shift Compensation (DSC) algorithm based on pilot tone

reliably capture frequency shifts due to platform velocities typical in acoustic localization scenarios. After bandpass filtering, the pilot tone is processed to determine its precise frequency shift. A common and effective technique for this step is the FFT. The filtered signal segment is transformed into the frequency domain, where the dominant frequency component reveals the pilot tone's Doppler-shifted frequency, f_r. By comparing f_r with the known original pilot frequency f_p, the Doppler frequency shift Δf is directly computed as: $\Delta f = f_r - f_p$.

(c) **Frequency Analysis**

An FFT of the filtered pilot segment identifies the peak frequency shift. The accuracy of frequency estimation via FFT depends on the duration of the observation window, as the frequency resolution of an FFT is inversely proportional to the time duration of the captured signal segment. To improve resolution without increasing the observation duration (and thus susceptibility to time-varying Doppler), zero-padding techniques are frequently employed. Zero-padding interpolates the frequency domain and provides a finer granularity of frequency estimation.

(d) **Frequency Correction**

Using the calculated frequency offset, the receiver adjusts the carrier frequency and potentially resamples the received waveform to correct time-scale distortions. Once the Doppler shift Δf is known, the next step involves generating a Doppler-compensated reference signal. The receiver adjusts the frequency of its locally generated reference to counteract the observed frequency offset. Specifically, the compensated carrier frequency $f_{c,new}$ is derived by:

$$f_{c,new} = f_c + \frac{f_c}{f_p} \Delta f \tag{4.4}$$

This correction aligns the reference waveform's frequency content precisely with the Doppler-affected received signal. A new reference signal will be generated using this compensated carrier frequency $f_{c,new}$.

(e) **Correlation and Demodulation**

Subsequently, cross-correlation between received signal and compensated reference signal is calculated to estimate the correct ToA.

This structured approach ensures that Doppler-induced errors, which could significantly degrade acoustic positioning accuracy, are effectively mitigated. Importantly, the computational requirements for this approach remain modest, particularly suitable for real-time operations on resource-limited platforms.

The pilot tone method for Doppler compensation offers notable advantages, including theoretical simplicity, computational efficiency, and demonstrated robustness in many practical implementations. By transmitting an auxiliary carrier tone to sense frequency shifts, the system can directly estimate relative velocity and correct the timing of the primary signal, which substantially improves the robustness and accuracy of acoustic positioning systems. However, this method also has significant limitations. It assumes uniform Doppler shifts across the frequency band, an assumption that may fail in multipath environments or when Doppler spreads vary. Pilot tones, being narrowband, are vulnerable to frequency-selective fading, and severe fading at the pilot frequency can degrade Doppler estimation accuracy. In addition, transmitting an extra tone consumes bandwidth and power, reduces the energy available for the main signal, and complicates multi-transmitter deployments, since each transmitter requires a distinct pilot frequency. On UAV platforms in particular, ambient noise and interference can further hinder the reliable detection of the pilot tone.

4.2.2 *Carrier Frequency Extraction via Square-Law Detection (Algorithm 2)*

Algorithm 2, based on carrier-frequency extraction, addresses Doppler shifts in a passive SSSLPS that uses FDMA (i.e., multiple beacons emitting distinct carrier frequencies simultaneously). The properties of the FDMA signals are shown in Table 4.1. The core idea is to extract the Doppler-shifted carrier frequency from the received spread-spectrum signal and use it to generate a frequency-corrected reference signal for correlation. By doing so, the correlation peak can be restored to full strength and the correct ToA determined despite the motion. This method assumes the receiver is moving while the transmit beacons remain fixed, which is a typical setup for robot localization in a known environment.

The DSC algorithm, illustrated in Fig. 4.5, consists of two main stages: estimating the Doppler-induced frequency shift and calculating the signal arrival time. The received signal $r(t)$ is first passed through a BPF to eliminate frequency components outside the main lobe, effectively reducing noise. The filtered signal is then multiplied by a sine wave to shift its spectrum to an intermediate frequency. The intermediate frequency used is f_{cI} = 4 kHz. The sine wave frequencies used for the four FDMA channels are 10, 18, 26, and 34 kHz, respectively. The resulting spectrum after this multiplication is shown in Fig. 4.6a.

To extract the Doppler-shifted carrier wave, the system uses the theory of a digital square-law detector. After frequency translation, the signal is passed through a low-pass filter (LPF) to remove high-frequency components. The resulting signal

Table 4.1 Properties of the FDMA signals

Channel No.	Carrier wave (kHz)	Chip rate (kcps)	BPF (kHz)	M_{length}
1	14	4	10~18	511
2	22	4	18~26	511
3	30	4	26~34	511
4	38	4	34~42	511

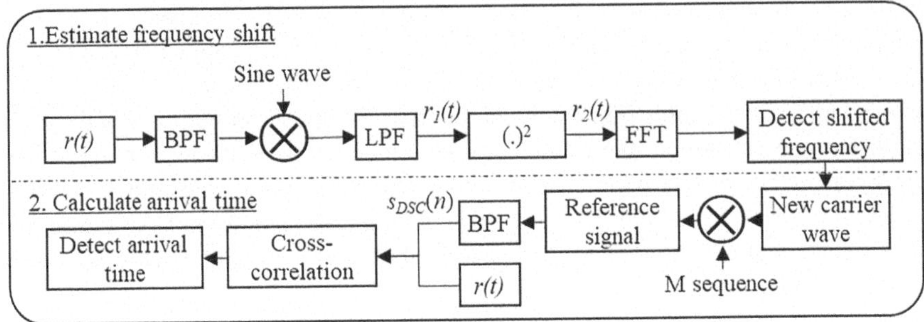

Fig. 4.5 Block diagram of the DSC algorithm based on carrier frequency detection

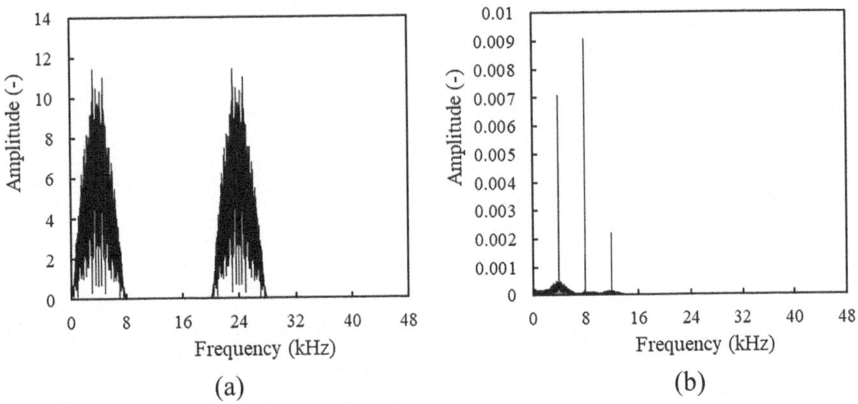

Fig. 4.6 An example of frequency spectrum after multiplying (**a**) sine wave and (**b**) squaring

$r_1(t)$ contains both the intermediate frequency component and the M-sequence modulation. It can be expressed as:

$$r_1(t) = A \times \sin\left(\frac{2\pi f_{c1}}{f_s} \times (t+\phi)\right) \times M\left(floor\left(\frac{f_{chipshift}}{f_s} \times (t-T_d)\right)\right) \tag{4.5}$$

where, f_{c1} is the Doppler-shifted carrier frequency, $f_{chipshift}$ is the Doppler-shifted chip rate, A is the signal amplitude, ϕ is the carrier phase shift, and T_d is the delay time of the M-sequence. Noise is neglected for clarity.

The signal $r_1(t)$ is then squared to remove the spreading code and highlight the frequency shift component, resulting in:

$$r_2(t) = A^2 \times M^2\left(floor\left(\frac{f_{chipshift}}{f_s} \times (t-T_d)\right)\right) \times \sin^2\left(\frac{2\pi f_{c1}}{f_s} \times (t+\phi)\right) \tag{4.6}$$

Since M^2 equals 1, Eq. (4.3) simplifies using a trigonometric identity to:

$$r_2(t) = \frac{A^2}{2} - \frac{A^2}{2}\cos\left(2 \times \frac{2\pi f_{c1}}{f_s} \times (t+\phi)\right) \tag{4.7}$$

The frequency spectrum of $r_2(t)$, shown in Fig. 4.6b, reveals a peak near 8 kHz. This peak, denoted p_{8k}, shifts in response to the Doppler effect. The peak frequency within the 7.9–8.1 kHz range corresponds to the Doppler shift and can be used to estimate the platform's velocity. For example, maximum measurable velocities are approximately 1.21, 0.77, 0.57, and 0.45 m/s for the four respective channels.

Using p_{8k}, a new Doppler-corrected carrier frequency and chip rate are computed:

$$\Delta f = \frac{p_{8k} - 2 \times f_{chip}}{2}$$

$$f_{cnew} = f_{cl} + \frac{p_{8k} - 2 \times f_{chip}}{2}$$

$$f_{chipnew} = f_{chip} + \frac{p_{8k} - 2 \times f_{chip}}{2} \tag{4.8}$$

A reference signal $s_{DSC}(l)$ is then generated using these new values:

$$s_{DSC}(l) = \sin\left(\frac{2\pi f_{cnew}}{f_s} \times l\right) \times M\left(floor\left(\frac{f_{chipnew}}{f_s} \times l\right)\right) \tag{4.9}$$

for $l = 0, 1, 2, \ldots, k_{DSC} - 1$, where k_{DSC} is the length of the reference signal with Doppler shift, as defined earlier.

One challenge is that the frequency spectrum may contain multiple peaks caused by reflections, which may distort the frequency shift estimate. To address this, the algorithm applies a threshold c_{fth} to identify the highest and second-highest peaks that are significantly above the noise floor. The threshold is defined as:

$$c_{fth} = 4\sigma_{fcorr}$$

$$\sigma_{fcorr} = \sqrt{\frac{\sum_{i=1975}^{2025} \left|f_{i\times 4} - \mu\right|^2}{51}} \tag{4.10}$$

where, σ_{fcorr} is the standard deviation of the frequency spectrum in the 7.9–8.1 kHz range, $f_{i\times 4}$ is the amplitude at frequency i × 4, and μ is the mean amplitude over that frequency range. The FFT is computed using 24,000 samples at a 96 kHz sampling rate, yielding a resolution of 4 Hz. If multiple peaks exceed the threshold, multiple candidate arrival times and distances are calculated. The algorithm then uses outlier rejection to select the most plausible range estimation.

Once the Doppler frequency shift Δf is estimated, the algorithm reconstructs a reference spread-spectrum signal that incorporates this shift. Specifically, the system generates a new local reference $s_{DSC}(n)$ using the same pseudorandom M-sequence but with the carrier frequency adjusted from f_c to $f_c^{'} = f_c + \Delta f$. In some implementations, the chip rate is also adjusted slightly if needed to maintain coherence with the Doppler-compressed or dilated signal. The length of this reference is set to cover the expected signal duration.

Finally, the corrected reference signal $s_{DSC}(n)$ is cross-correlated with the received signal $r(t)$. Because the reference now accounts for the motion-induced frequency shift, the cross-correlation yields a strong, distinct peak at the true time-of-flight. Standard peak-detection logic can then extract the ToA from the correlation. Any residual timing bias due to higher-order effects (e.g., acceleration during the signal's travel) is generally very small if Δf is accurately estimated, and can be neglected or handled by higher-level filtering.

In cluttered indoor environments such as greenhouses or laboratories, multipath propagation can occur when a microphone receives both the direct sound signal and one or more reflected signals from surrounding surfaces. These reflected paths often introduce slightly different Doppler shifts compared to the direct path because of differences in travel distance and angle. As a result, applying the same Doppler compensation to all received signals can degrade the accuracy of the correlation result and lead to incorrect time of arrival estimation. This effect is illustrated in Fig. 4.4, where Fig. 4.7a shows a strong and sharp correlation peak after correct compensation for the direct signal, while Fig. 4.7b shows a broader and weaker peak due to mismatched compensation for the reflected signal. To address this issue, Huang et al. proposed a method that first detects multiple significant peaks in the frequency domain, each representing a possible Doppler shift. The algorithm then generates corresponding reference signals for these candidate frequencies and performs correlation with each. The final distance estimate is selected using an outlier rejection process that favors the correlation result with the highest signal to noise ratio, which typically corresponds to the direct path. This strategy enhances the

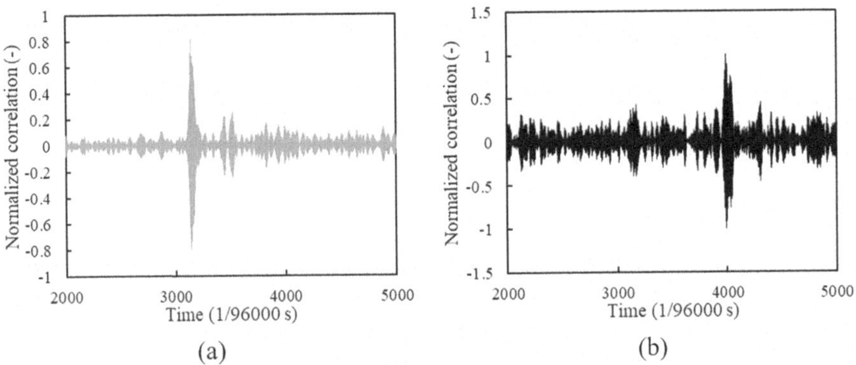

Fig. 4.7 Correlation after compensation of (**a**) direct sound wave and (**b**) reflected sound wave

robustness of the positioning system in environments where reflections are unavoidable.

In scenarios with complex motion (e.g., a robot turning or accelerating so that its relative velocity to each beacon changes during the signal's propagation), the frequency-compensated reference may not perfectly match the received signal, causing a slight reduction in correlation peak clarity. This can lead to occasional small outliers in the range estimates. Huang et al. observed a few cases of approximately 10 cm error when velocity changes were significant. They suggested further improvements, such as investigating demodulation techniques to remove the carrier entirely or integrating inertial sensors to predict Doppler shifts during rapid maneuvers. Nonetheless, the carrier-frequency extraction DSC proved to be a robust solution for moderate speeds and it formed the foundation for subsequent research on Doppler compensation.

4.2.3 Multi-candidate Spectrum Peak Correlation (Algorithm 3)

The algorithm 3 was proposed by Tsay et al. (2023), for localizing a dynamic UAV with SSSLPS. It tackles Doppler compensation via a brute-force search in the frequency domain (shown in Fig. 4.8). Instead of directly extracting the Doppler shift from the signal (as in Algorithm 1), it generates a set of reference signals with various assumed Doppler shifts and finds which one best correlates with the received signal. This approach is especially useful when the platform's motion might be more complex or when the signal-to-noise ratio is lower (as is often the case for UAVs due to motor noise and fast movement). The authors call this a "spectrum peak algorithm" because it effectively searches for the reference whose correlation produces the highest peak (highest SNR). It leverages prior knowledge of the UAV's approximate speed range to constrain the search, making it computationally feasible.

Before processing each acoustic frame, the algorithm estimates a plausible frequency shift range based on the UAV's expected speed. For instance, in the target scenario the UAV was intended to cruise at ~0.2 m/s for monitoring crops in a greenhouse. Using the Doppler formula, a speed of 0.2 m/s with sound speed ~343 m/s and carrier frequency ~24 kHz yields a maximum shift of $\Delta f \approx \pm 15$ Hz. Thus, the algorithm limits its search to a band of frequencies around the nominal

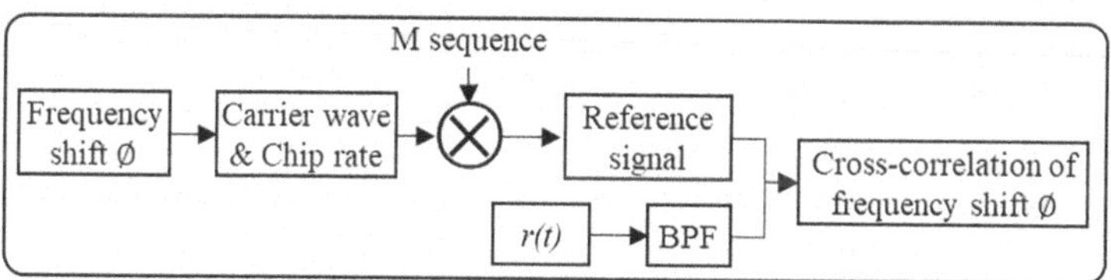

Fig. 4.8 Block diagram of the multi-candidate spectrum peak correlation DSC algorithm

carrier. In their implementation, they considered offsets from −15 Hz to +15 Hz (relative to the base 24 kHz carrier) in 1 Hz increments. This yields 31 candidate frequency shifts, denoted by $\Theta = \{-15, -14, \ldots, 0, \ldots, +14, +15\}$ Hz.

For each candidate shift $\theta \in \Theta$, a Doppler-adjusted reference signal is generated: the carrier frequency is set to $f_c + \theta$, and importantly, the spreading code chip rate is proportionally adjusted as well. Adjusting the chip rate (slightly) ensures that the reference signal's time axis is effectively scaled to account for the relative motion. This models the fact that if the UAV is moving toward a beacon, the incoming chips are slightly compressed in time, and if moving away they are slightly stretched. In practice, Tsay et al. defined an updated chip frequency as $f_{chip} = (f_c + \theta)/2$ (since their original chip rate was half the carrier frequency for BPSK modulation). Each reference signal covers the duration of one ranging cycle (e.g., one TDMA slot) and uses the known M-sequence for that beacon's channel.

Next, the received signal (for the current beacon's time slot) is correlated against each candidate reference signal. That is, 31 cross-correlation computations are performed, yielding 31 correlation waveforms. The algorithm then selects the reference signal that yielded the highest SNR_{corr}. The intuition is that the correctly Doppler-compensated reference will align best with the received signal, producing a strong, sharp peak, whereas references with the wrong frequency offset will produce a smeared correlation with lower peak SNR. By picking the best among the candidates, the algorithm simultaneously identifies the Doppler shift (the θ that was used for that reference) and obtains the precise ToA from the corresponding correlation peak.

Figure 4.9 shows an example where the algorithm tests several possible Doppler shifts to determine which one best matches the received signal. In this case, the original signal has a carrier frequency of 24 kHz, and the system evaluates shifts between −15 and 15 Hz. For each candidate shift, it calculates how well the adjusted reference signal aligns with the received signal by measuring the SNR_{corr}. In the example, only one of the tested frequency shifts produces an SNR_{corr} greater than 18 dB, showing that it is a strong match and likely the correct shift. All the other tested shifts result in lower SNR_{corr} values, meaning they do not match the received signal as well. This demonstrates how the system selects the correct Doppler shift by identifying the candidate that gives the highest correlation strength.

In essence, this approach performs a direct search for the Doppler shift: rather than explicitly computing an FFT to find the frequency offset, it tries a series of small frequency adjustments and finds the one that maximizes the correlation peak. This method is computationally heavier than Algorithm 2 (since many correlations are computed), but with a modern processor or Field-Programmable Gate Array (FPGA) the load is manageable for a moderate number of candidates (31 in this case is quite small). Moreover, the search range can be tailored to the platform's expected maximum speed. For faster-moving platforms, the range or step size could be widened, trading more computation for the ability to handle larger Doppler shifts.

After selecting the best-match reference signal, the ToA is determined by the index of the peak in that correlation and distance from speaker to microphone can

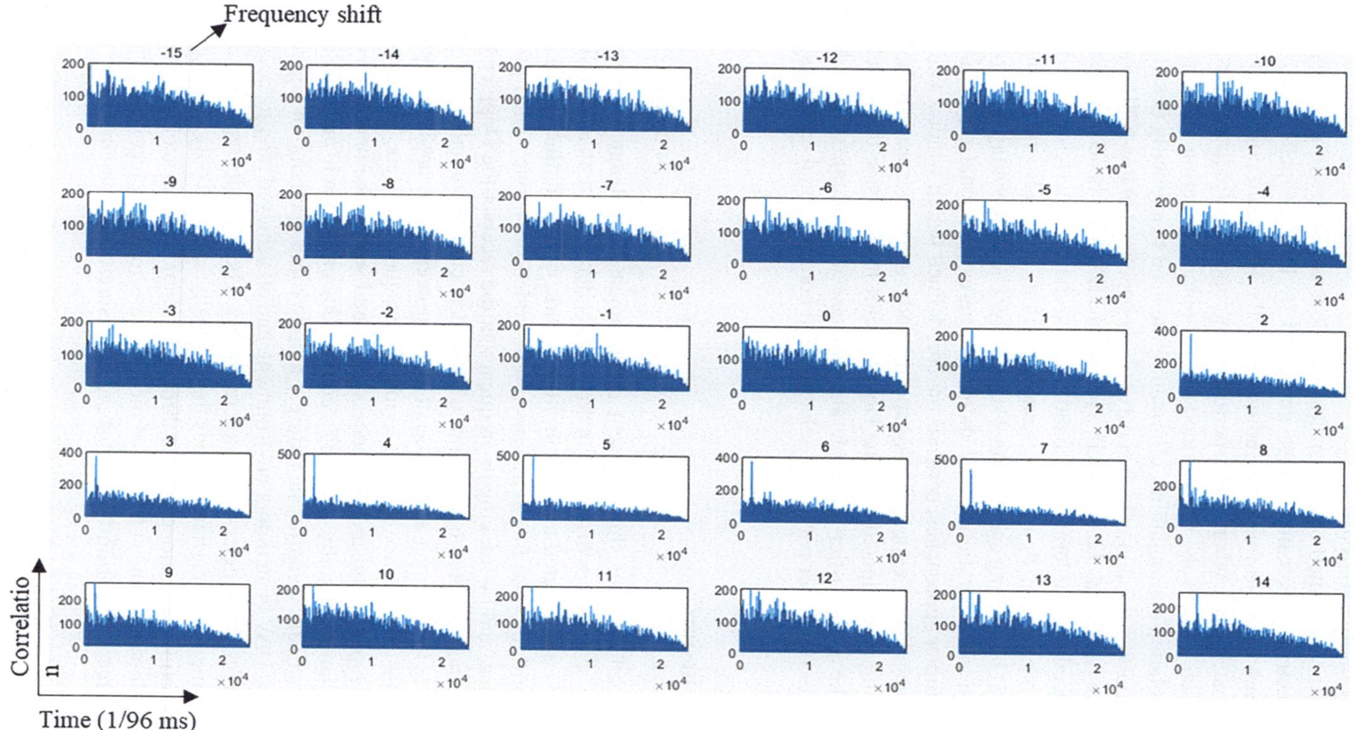

Fig. 4.9 Example of correlation results at different assumed frequency shifts

be calculated. Finally, with multiple beacons in the system, the position of the UAV is obtained by multilateration.

The spectrum peak method demonstrated a clear advantage in its robustness to the broad-spectrum mechanical noise generated by UAV propellers and motors, since by testing multiple frequency adjustments the algorithm could still identify the correct match even when noise perturbed the apparent frequency. Its search-based approach effectively transformed the Doppler estimation problem into a series of correlation computations, eliminating the need to explicitly detect the carrier in the frequency domain, which is particularly beneficial in low-SNR conditions where spectral peaks may be obscured. The use of TDMA for beacon transmissions further enhanced noise tolerance by avoiding inter-channel interference and simplifying signal processing, though the same algorithm could also be applied in an FDMA context by treating each channel separately. The disadvantages are primarily computational and operational. The method requires additional processing compared with simpler approaches, depends on a predefined search window Θ that must be carefully matched to expected UAV speeds, and can lose accuracy if the vehicle moves faster than anticipated or if the Doppler shift falls outside the search range. Thus, while the method offers high tolerance to noise and dynamic motion, its effectiveness hinges on proper parameter selection and computational resources.

4.3 Conclusion

Motion-induced Doppler shifts present a fundamental challenge to sound-based positioning. The advances discussed in this chapter demonstrate that these shifts can be effectively compensated to maintain high accuracy. We examined three complementary algorithms. The first relies on an embedded pilot tone which provides a direct reference, and by tracking its frequency at the receiver the Doppler shift can be estimated and corrected in real time with modest computational effort. The second determines the Doppler shift through signal processing by applying square-law detection to the carrier wave in order to identify frequency offsets. The third generates several Doppler-adjusted reference signals and selects the one that yields the highest correlation, thereby ensuring accurate detection. Each method has been validated in realistic scenarios including a mobile robot operating in a greenhouse and a dynamic UAV in an indoor environment, and all achieved ranging accuracy on the order of a few centimeters with high reliability. These results represent a significant improvement over earlier methods that typically achieved only 20–30 cm accuracy under motion. Taken together, the three algorithms allow an acoustic positioning system to follow frequency changes of moving transmitters or receivers in real time and ensure that correlation-based distance measurements remain sharp, reliable, and precise.

The choice of algorithm depends on application needs. The pilot tone method (Algorithm 1) is simple and efficient, providing real-time Doppler estimates from a reference tone but sensitive to fading at the pilot frequency. The carrier-extraction method (Algorithm 2) is also lightweight and effective for FDMA systems with moderate velocities. The multi-candidate correlation method (Algorithm 3) requires more computation yet offers strong robustness under low SNR and for UAVs with fast or erratic motion, achieving more than 90% detection success and an average error of about 7.6 cm. All three methods substantially reduced missed detections and improved accuracy by a factor of three to four compared with uncompensated cases. Looking forward, their performance can be enhanced by combining with inertial sensors, predictive filters, and adaptive codes, while future research should also address Doppler spread. Together, these techniques demonstrate that SSSLPS with Doppler compensation can deliver centimeter-level accuracy for mobile robots and UAVs in GPS-denied environments, enabling reliable indoor navigation in farms, warehouses, and factories.

References

Huang, Z., Shiigi, T., Tsay, L. W. J., Nakanishi, H., Suzuki, T., Ogawa, Y., & Naoshi, K. (2021). A sound-based positioning system with centimeter accuracy for mobile robots in a greenhouse using frequency shift compensation. *Computers and Electronics in Agriculture, 187*, 106235. https://doi.org/10.1016/j.compag.2021.106235

Ko, S., Choi, J., & Kim, B. (2008). Indoor mobile localization system and stabilization of localization performance using pre-filtering. *International Journal of Control, Automation, and Systems, 6*(2), 2.

Smith, A., Balakrishnan, H., Goraczko, M., & Priyantha, N. (2004). Tracking moving devices with the cricket location system. In *Proceedings of the 2nd international conference on Mobile systems, applications, and services - MobiSYS '04* (p. 190). https://doi.org/10.1145/990064.990088

Tsay, L. W. J., Tomoo, S., Huang, Z., Nakanishi, H., Suzuki, T., Shiraga, K., Ogawa, Y., & Kondo, N. (2023). An acoustic based local positioning system for dynamic UAV in GPS-denied environments. *Applied Engineering in Agriculture, 39*(3), 3. https://doi.org/10.13031/aea.15397

Widodo, S., Shiigi, T., Hayashi, N., Kikuchi, H., Yanagida, K., Nakatsuchi, Y., Ogawa, Y., & Kondo, N. (2013). Moving object localization using sound-based positioning system with Doppler shift compensation. *Robotics, 2*(2), 2. https://doi.org/10.3390/robotics2020036

Chapter 5
Environmental Effects and Compensation Strategies in Acoustic Positioning

Acoustic positioning systems, while highly effective in controlled indoor settings, can experience significant accuracy degradation under varying environmental conditions. Factors such as temperature fluctuations and wind can alter the propagation speed and path of sound, introducing errors in distance measurement and localization. For instance, a mobile robot operating in a greenhouse or an open field may encounter temperature gradients or wind gusts that were not present in the laboratory, leading to systematic timing errors if not properly accounted for.

This chapter examines how environmental conditions affect acoustic positioning and presents methods to mitigate these effects. Section 5.1 analyzes the influence of temperature and wind on sound propagation and positioning accuracy, providing theoretical background. Section 5.2 then discusses compensation strategies to correct for these environmental distortions. In particular, two methods developed to address temperature and wind effects are described: one algorithm estimates the sound velocity in real-time, and another uses a base station approach to interpolate wind-induced velocity changes across a field. Finally, Section 5.3 offers a discussion on the trade-offs between different compensation approaches and gives design recommendations for deploying acoustic positioning systems in variable environmental conditions. The emphasis is on conceptual understanding and algorithmic solutions, without delving into the details of experimental setups, so that the reader can apply these principles to a range of scenarios.

5.1 Effects of Environmental Conditions on Positioning Accuracy

Environmental factors can significantly influence the accuracy of acoustic time-of-flight measurements by altering the speed and path of sound waves. The most pertinent conditions are temperature and wind, especially in semi-indoor environments

Z. Huang, *Acoustic and Signal-Based Local Positioning*, Navigation: Science and Technology 18, https://doi.org/10.1007/978-981-95-6083-7_5

like greenhouses and in fully outdoor open-field environments. Understanding their effects on sound propagation is crucial for diagnosing errors and motivating compensation techniques.

5.1.1 *Temperature Effect on Sound-Based Positioning System*

Temperature is one of the primary factors determining the speed of sound in air. In general, sound travels faster in warmer air and slower in cooler air. The ideal gas law relationship for sound speed (assuming dry air) can be approximated by a linear formula over typical temperature ranges. At 0 °C (273 K) the speed of sound is about 331 m/s, and it increases by roughly 0.6 m/s for every 1 °C increase in air temperature. In equation form, a useful approximation is:

$$v_i = 331.5 + 0.61\left(T_{si}\right) \tag{5.1}$$

where, T_{si} is the air temperature in °C and v_i is the speed of sound in m/s. Thus, at a warm temperature of 20 °C (typical room temperature), the speed of sound is about 343.4 m/s = 331.3 + 0.606 × 20.

Temperature-induced changes in sound speed are substantial. An increase of 10 °C results in roughly a 3.5% higher sound speed, whereas a drop in temperature slows sound by a similar percentage. In an acoustic positioning system, which often assumes a nominal sound speed (e.g. 343 m/s at 20 °C), any deviation in actual air temperature will cause a proportional range error. For example, if the air cools to 10 °C (c ≈ 337 m/s) but the system still assumes 343 m/s, a distance measurement of 10 m based on ToF would be underestimated by nearly 2%. This corresponds to an error of about 20 cm, which is far greater than the centimeter-level accuracy goal. Clearly, unmodeled temperature variation can be a dominant error source.

Temperature is rarely uniform in real environments. Greenhouses in particular exhibit spatial and temporal temperature gradients. Solar heating, plant transpiration, and ventilation create microclimates where air temperature can vary by several degrees between the floor and the ceiling or between sunlit and shaded areas. For instance, one study observed up to an 11 °C temperature difference within a small greenhouse from morning to afternoon, resulting in measurable shifts in acoustic positioning accuracy (Fig. 5.1) (Tsay et al., 2020). Warm air tends to rise, forming layers of different temperature especially in winter when heating at ground level and a cold roof create stratification. These temperature layers can bend sound waves (an effect known as refraction, where sound curves toward cooler, slower propagation regions) and can also cause varying delays if a robot moves between zones of different temperature. In open fields, temperature is generally more homogeneous across horizontal distances, but there is often a vertical gradient (e.g. ground heating on a sunny day can produce warmer air near the surface). Rapid temperature changes over time (day vs. night) can also be an issue; however, open fields typically have air mixing that reduces sharp spatial gradients compared to enclosed spaces. In any

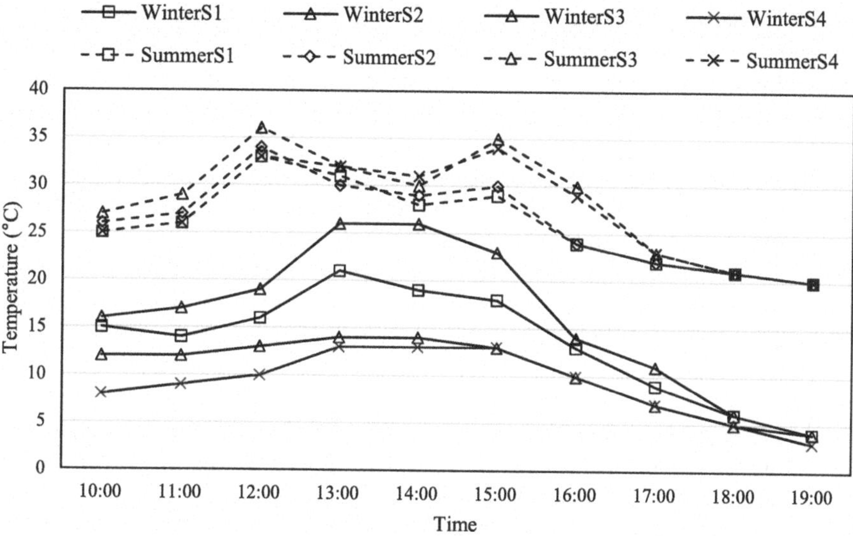

Fig. 5.1 Temperature change in a greenhouse during a day

case, if an acoustic positioning system does not continuously adjust for the current effective sound speed, even modest temperature changes can degrade its accuracy.

5.1.2 Wind Effect on Sound-Based Positioning System

Wind is a critical environmental factor in open fields and large open indoor spaces, introducing anisotropy (directional differences) in sound propagation speed and travel time. Wind can be thought of as the movement of the medium (air) through which sound waves travel. If air is moving, it effectively carries the sound waves with it in the direction of the wind. From the perspective of an observer on the ground, a sound traveling downwind (with the wind) will appear to travel faster, while sound traveling upwind (against the wind) travels slower than it would in still air. For a wind blowing at speed v_{wind} in a certain direction, the effective sound velocity v_{eff} along that direction is approximately the vector sum of the still-air sound speed and the wind velocity component in that direction. In mathematical terms, if θ is the angle between the sound propagation direction and the wind direction (with $\theta = 0$ meaning sound propagates exactly downwind and $\theta = 180°$ upwind), then:

$$v_{eff} = v + v_{wind} Cos\theta \tag{5.2}$$

where v is the sound speed in still air. Eq. (5.2) implies that for $\theta = 0$ (sound and wind aligned), $v_{eff} = v + v_{wind}$ (sound speed is increased by the wind speed), and for

$\theta = 180°$ (opposing the wind), $v_{eff} = v - v_{wind}$ (sound is effectively slowed down). If sound travels perpendicular to the wind ($\theta = 90$ or $270°$), $Cos\theta = 0$ and wind has minimal first-order effect on propagation speed (though turbulence can still affect the path slightly).

The directional propagation delay caused by wind translates into measurable timing asymmetry. Consider an acoustic positioning system in an open field with one beacon to the north and another to the south of a robot. If a steady wind blows from north to south, sound traveling from the north beacon to the robot (downwind) will arrive sooner than expected (relative to no wind), whereas sound from the south beacon (upwind) will arrive later. If uncorrected, the localization algorithm will interpret these time-of-flight differences as the robot being closer to the north beacon and farther from the south than it really is, skewing the computed position. Even moderate winds can cause significant errors: a 5 m/s wind (about 18 km/h) is roughly 1.5% of the nominal sound speed (343 m/s). Over a 30 m distance, a fully downwind sound path would arrive as if the distance were shorter by ~0.45 m (1.5% of 30 m), which is a huge error for a system trying to achieve, say, 20 mm accuracy. Wind gusts or turbulent airflows add further variability, causing the sound speed to fluctuate over time and space. In greenhouse environments, strong winds are usually absent (since they are enclosed or semi-enclosed spaces), but there may be gentle air currents from ventilation fans or convection. These flows are typically much slower (on the order of 0.1–1 m/s) and less uniform than outdoor wind, so their effect on acoustic propagation is smaller and more localized. Nonetheless, any systematic airflow in an indoor farm or large building (for example, airflow in long corridors or large halls) could introduce slight bias in one direction of acoustic measurements if not accounted for.

In summary, environmental conditions directly impact the accuracy of acoustic positioning by altering the speed and trajectory of sound. Temperature changes lead to sound speed estimation errors if a fixed value is assumed, and spatial temperature gradients can cause non-uniform propagation delays across a workspace. Wind creates direction-dependent time-of-flight errors, effectively stretching or compressing measured distances along the wind axis. Both factors can easily produce errors on the order of several centimeters or more if left uncompensated. These challenges necessitate compensation strategies, which we will explore next, to maintain high positioning accuracy in variable environments.

5.2 Compensation Strategies for Environmental Influences

To ensure reliable acoustic localization in the face of temperature and wind variations, various compensation strategies can be employed. The general principle behind all such strategies is to either measure and correct for environmental effects in real time, or to design the localization algorithm to be inherently robust against those effects. Some broad approaches include adjusting the assumed sound speed based on temperature sensors, filtering out wind-induced errors using additional data, and simultaneously estimating environmental parameters along with position. In essence, the system must continuously calibrate itself to the current environment. The following are key principles for correcting environmental distortion.

One effective approach is to continuously update the sound propagation speed used in distance calculations. This can be done by reading physical sensors (e.g. thermometers for temperature, anemometers for wind) or by algorithmically estimating the effective sound speed from the acoustic signals themselves. Figure 5.2a shows two sensor that could be used for measure temperature, where is put into a wind tube to get temperature data accurately. Figure 5.2b is a wind meter which can provide wind velocity and direction information.

Another effective strategy is to incorporate environmental parameters, such as sound speed or wind velocity, as unknown variables within the TOA positioning algorithm. By using a sufficient number of reference nodes or additional signals, the system can solve for these parameters concurrently with the target's position, eliminating the need for external sensors.

In addition, fixed reference anchors or base stations can be used to periodically transmit or receive signals with known positions and timings. By analyzing the time-of-flight of these signals and comparing them to expected values based on known distances, the system can detect environmental deviations. For instance, a signal arriving later than expected may indicate cooler air or the presence of a headwind. These deviations can then be used to adjust measurements across the rest of the workspace.

In environments that are large or have varying environmental conditions, it may be advantageous to map the spatial distribution of temperature or wind. This can be done using sensor networks or distributed acoustic measurements. The workspace can then be divided into zones with relatively uniform environmental properties,

Fig. 5.2 Sensors for estimating (**a**) temperature and (**b**) wind

and each zone can be assigned specific sound speed values or correction factors to improve positioning accuracy.

Implementing these principles can greatly reduce errors. In the following, we describe two specific compensation methods developed to address temperature and wind effects, respectively. The first method uses a TOA algorithm to estimate the sound velocity in real-time as part of the localization solution, thereby compensating for temperature-induced speed changes without any physical temperature sensors. The second method introduces a base station technique to dynamically correct for wind by measuring its effect on acoustic signals and interpolating this across the field.

5.2.1 *Temperature Compensation via Sound Velocity Estimation*

One effective strategy to handle temperature fluctuations is to make the speed of sound itself a state variable in the localization algorithm. Instead of assuming a fixed velocity or relying on external thermometers, the system can solve for the robot's position and the effective sound speed simultaneously using time-of-arrival measurements. This approach essentially treats the average sound velocity in the environment as an unknown to be estimated at each localization update. By doing so, any changes in temperature (which directly influence sound speed per Eq. (5.1)) are implicitly accounted for in the solution, a form of real-time temperature compensation.

The proposed method using an estimated sound velocity for calculating the position and the sound velocity is based on a TOA localization algorithm. Eq. (5.5) indicates the relationship between distances and coordinates of each node and measurement position simultaneously. This differs from the conventional method (Eq. 5.4), which uses data from temperature sensors.

$$(t_i)(v_e) = \sqrt{(x_s - x_i)^2 + (y_s - y_i)^2 + (z_s - z_i)^2}$$

$$f_i(x_s, y_s, z_s, v_e) = \sqrt{(x_{s0} - x_i)^2 + (y_{s0} - y_i)^2 + (z_{s0} - z_i)^2} - (t_i)(v_e) \quad (5.3)$$

where, t_i is propagation time of emitted sound of the i_{th} node, (x_s, y_s, z_s) is the estimated position of the target, (x_i, y_i, z_i) is the position of the ith node, and v_e is the estimated sound velocity. It is assumed in Eq. (5.3) that the sound velocity of the propagated sound between the transmitters and receiver in the greenhouse is constant.

The unknowns are the position coordinates (x_s, y_s, z_s) and the estimated sound velocity v_e. When the transmitting time t_i is measured, the positions of the nodes are known and the provisional sound velocity can be set. Therefore, the four unknowns (x_s, y_s, z_s) and v_e can be estimated when the number of nodes is larger than four, since at least a system of four equations in Eq. (5.3) are needed in order to solve the four unknowns. The function, $f_i(x_s, y_s, z_s, v_e)$, represents the positioning error, which indicates the distance of the target position minus the sound propagation distance.

This is difficult to compute because Eq. (5.3) is a non-linear equation. Thus, it needs to be linearized by a Taylor expansion, (x_s, y_s, z_s) and v_e estimated by sequential computation of the function $f_i(x_s, y_s, z_s, v_e)$, which is also defined in Eq. (5.3). $f_i(x_s, y_s, z_s, v_e)$ is linearized by the first-order Taylor-series expansion at $x_{s0}, y_{s0}, z_{s0}, v_{e0}$, as in Eq. (5.4).

$$\begin{aligned} f_i(x_s, y_s, z_s, v_e) = & \sqrt{(x_{s0} - x_i)^2 + (y_{s0} - y_i)^2 + (z_{s0} - z_i)^2} - (t_i) v_{e0} \\ & + \frac{\partial f_i}{\partial x_s}(x_s - x_{s0}) + \frac{\partial f_i}{\partial y_s}(y_s - y_{s0}) + \frac{\partial f_i}{\partial z_s}(z_s - z_{s0}) \\ & + \frac{\partial f_i}{\partial v_e}(v_e - v_{e0}) \end{aligned} \tag{5.4}$$

The defined matrixes and vectors are as in Eq. (5.7)

$$\Delta d_i = \frac{\partial f_i}{\partial x_s}\Delta x_s + \frac{\partial f_i}{\partial y_s}\Delta y_s + \frac{\partial f_i}{\partial z_s}\Delta z_s + \frac{\partial f_i}{\partial v_e}\Delta v_e$$

$$\Delta x_s = (x_s - x_{s0})$$

$$\Delta y_s = (y_s - y_{s0})$$

$$\Delta z_s = (z_s - z_{s0})$$

$$\Delta v_e = (v_e - v_{e0})$$

$$\Delta d = \begin{bmatrix} \Delta d_2 \\ \Delta d_3 \\ \vdots \end{bmatrix} \mathrm{A} = \begin{bmatrix} \frac{\partial f_2}{\partial x_s} & \frac{\partial f_2}{\partial y_s} & \frac{\partial f_2}{\partial z_s} & \frac{\partial f_2}{\partial v_e} \\ \frac{\partial f_3}{\partial x_s} & \frac{\partial f_3}{\partial y_s} & \frac{\partial f_3}{\partial z_s} & \frac{\partial f_3}{\partial v_e} \\ \vdots & \vdots & \vdots & \vdots \end{bmatrix}$$

$$\Delta x = [\Delta x_s \;\; \Delta y_s \;\; \Delta z_s \;\; \Delta v_e]^T$$

$$\Delta d = A\Delta x \tag{5.5}$$

A is the observation matrix. The generalized inverse matrix of A is multiplied on both sides of the Eq. (5.5). An iterative least squares method is used by iterating 50 times when Δx is approaching 0 and gives the approximate coordinate of the target.

$$\Delta x = \left(A^T A\right)^{-1} A^T \Delta d \tag{5.6}$$

Effectively, the algorithm is self-calibrating the sound speed using the distances between the target and multiple anchors. Intuitively, if the initial guess of v_{e0} is too high or too low, the computed distances will not consistently intersect at one point for all anchors. The solver then adjusts v_e until a consistent solution is found, which corresponds to using the "correct" average sound speed for that snapshot in time.

The advantages of this approach are significant in environments with spatial or temporal temperature variation. Since the algorithm finds the average sound velocity along the propagation paths, it accounts for the actual air conditions without needing to deploy temperature sensors throughout the environment. This was demonstrated by Tsay et al. (2020) in a greenhouse scenario, where their spread-spectrum acoustic positioning system was augmented with an on-line sound speed estimation algorithm. The system achieved high accuracy across different seasons by constantly updating v_e as part of the localization calculation. Notably, their results showed a positioning error below 20 mm in both summer and winter, even though internal greenhouse temperatures varied widely (by more than 10 °C). This accuracy was actually better than that of a traditional approach based on temperature sensor readings. Under the same conditions, the sensor-based compensation produced errors on the order of 30 mm. The self-estimation method outperformed it because it effectively captures the effective sound speed along the sound paths, whereas a few discrete temperature sensors might not reflect the true average temperature of the entire path (for example, sensors could be placed at locations that are slightly cooler or warmer than the path of the acoustic signal).

There are, however, important considerations when using this method. First, it requires more infrastructure and computational complexity to solve the expanded localization problem. Tsay et al. note that the calculation is more involved and that having sufficient anchor nodes is critical. Second, the method assumes that the sound speed is roughly uniform over the scale of the environment (or at least along the paths from the target to each anchor). In a small greenhouse or indoor space, this assumption holds well, and the algorithm identifies a single average temperature for the space. If there were extreme heterogeneity (e.g. a sharply localized hot spot causing a big speed gradient), a single v_e might not perfectly linearize all distances, potentially leaving some residual error. In practice, for the scenarios tested (small greenhouse with some stratification), a single estimated sound speed per update was sufficient to compensate most of the error. The success of this approach demonstrates that real-time sound velocity estimation is a powerful way to achieve temperature compensation. By embedding the compensation into the positioning algorithm itself, the system attains a form of adaptive calibration, always using the appropriate speed of sound for the current conditions without requiring manual intervention. This approach is especially valuable in indoor robots operating in environments like greenhouses, factories, or large indoor arenas where temperature can drift or vary across space.

5.2.2 Wind Compensation Using a Base Station Interpolation

In outdoor or open environments where wind is a dominant factor, a different strategy is needed. Unlike temperature, which primarily affects the scalar value of sound speed uniformly in all directions, wind introduces a vector field whose impact on acoustic propagation depends on direction, as described in Sect. 5.1. A wind compensation method using a base station configuration has proven effective for open-field acoustic positioning. The core idea is to use a dedicated reference emitter (or receiver) at a known location, the base station (BS), to continuously probe the environment and measure how wind is altering sound travel times in various directions. These measurements are then used to interpolate or estimate the sound-speed adjustments for arbitrary paths in the field, allowing correction of the robot's position calculations to account for the wind. In this conceptual diagram, a central BS emits acoustic signals to multiple receivers around the field. A uniform wind (arrow indicating wind direction) causes the sound to travel faster toward downwind anchors and slower toward upwind anchors. By measuring these time-of-flight differences at the anchors, the system interpolates the wind's effect across the area and compensates for it in the robot's localization.

In one implementation of this concept (Widodo et al., 2014), an inverse-GPS style configuration was employed for a spread-spectrum sound positioning system. Four microphones placed at the corners of a 30 m × 30 m area acted as fixed receivers, and an omnidirectional speaker at the base station broadcast acoustic signals. In essence, the base station played the role of a "satellite" on the ground, sending out a signal that all anchors receive. In calm air, the time-of-flight from the base station to each anchor can be predicted from their known distances and a known sound speed. However, when wind is present, those travel times will differ: signals traveling downwind from the base station arrive sooner than expected, and upwind signals arrive later. By comparing the measured travel times to the expected times (with no wind), the system can deduce the effective sound speeds in the directions toward each anchor. From two or more such directions, an approximation of the wind's effect (its speed and direction) can be derived. The setup of this wind compensation system using a base station is shown in Fig. 5.3.

A simple scenario helps to illustrate this point. Suppose the base station is at the center of the field with anchors placed to the north and south at equal distances. If a steady wind blows from north to south, the time to the south anchor will be shorter than the time to the equally distant north anchor. The system could compute an estimated wind speed component from this time difference. More generally, with anchors around the base station, it is possible to solve for a wind velocity vector W that best explains the pattern of time-of-flight deviations observed. This process is analogous to determining wind speed and direction by observing how the propagation of sound is pushed one way or another. Mathematically, if $W = (W_x, W_y)$ is the horizontal wind vector and u_i is the unit vector from the base station to anchor i, the measured time-of-flight t_i to anchor i can be related to the wind as:

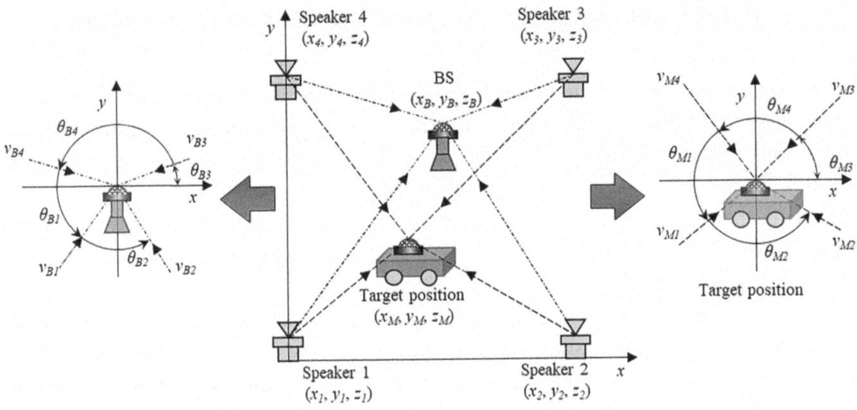

Fig. 5.3 Setup of the wind compensation system using a base station

$$t_i \approx \frac{d_i}{v + W \cdot u_i} \tag{5.7}$$

where $d_i = |a_i|$ is the distance from the base station to the ith anchor (with a_i the position vector of that anchor relative to the base station) and v is the sound speed in still air (accounting for temperature as needed). For small wind speeds relative to v, we can linearize this relation. Rearranging the above (and neglecting second-order terms) yields approximately:

$$\mathrm{W} \cdot u_i \approx t_i d_i - v \tag{5.8}$$

In other words, the component of the wind along the ith path is equal to the difference between the effective sound speed along that path (d_i/t_i) and the nominal sound speed v. With two or more anchors at different directions, these Eq. (5.5) can be solved to find the wind vector W. For instance, with two anchors in roughly opposite directions (e.g., north and south), one can solve two linear equations for W_x and W_y directly. With more anchors, an over-determined set of equations can be solved in a least-squares sense to obtain the wind estimate that best fits all the measurements.

Once W is estimated, any acoustic range measurement in the field can be corrected by projecting W onto that measurement's propagation path (as per Eq. (5.2)). Effectively, the algorithm adjusts the assumed travel time by subtracting the wind's along-path contribution. This technique assumes the wind is relatively uniform across the area, which is a reasonable approximation for moderate winds over tens of meters in open space without large obstacles.

The base station method has been shown to greatly reduce positioning errors due to wind. With the base station located centrally in a 30 m field, the wind-compensated acoustic system achieved an average 2D localization error of about 20 mm at an update rate of 2 Hz, whereas without compensation, the same system's error in that wind condition was around 70 mm on average. In other words, the wind-induced error was cut by more than a factor of three by using the base station calibration, which brought performance back to the few-centimeter level even under nontrivial wind conditions.

The base station wind compensation method shows that placement is critical, since positioning the base station at the center provides balanced calibration directions and reliable interpolation under uniform winds, whereas placing it off to one side limits effectiveness and can even lead to overcompensation. The approach works well when wind can be approximated by a single linear vector, such as a steady breeze, but becomes less accurate under complex conditions like turbulence or varying wind speeds across sub-areas; in such cases, multiple base stations, higher-order models, or integration with direct wind sensor readings may be needed. Although the method introduces some overhead by dedicating reference signals for calibration, the cost is modest and yields a robust real-time estimate of wind impact, which is invaluable for maintaining centimeter-level accuracy in outdoor or large-scale acoustic positioning applications.

5.3 Discussion and Design Recommendations

The above compensation strategies highlight a fundamental trade-off in designing acoustic positioning systems for variable environments, namely the balance between complexity and robustness. Different methods offer different advantages, and choosing the right approach (or combination of approaches) depends on the expected environmental variability, the required accuracy, and the resources (hardware and computational) available. Here we discuss trade-offs and provide recommendations for practitioners and designers.

A straightforward way to counteract environmental effects is to use physical sensors and feed their readings into the acoustic ranging calculations. This is conceptually simple and may work well if conditions are uniform or change slowly. However, sensor-based compensation can struggle to capture localized or fast-changing phenomena—e.g., a temperature sensor in one corner of a greenhouse might not notice a pocket of warm air in another corner, or a single anemometer might not represent the wind at all points in a large field. The algorithmic compensation methods (like the sound-speed estimation of Sect. 5.2.1 and the base station interpolation of Sect. 5.2.2) directly leverage the acoustic signals themselves to infer the environmental effect on the fly. These methods tend to provide more accurate and locally relevant corrections because they measure what actually matters for positioning, rather than relying on separate ancillary sensors. The trade-off is increased complexity: for example, solving simultaneously for position and sound speed means needing additional anchor nodes and a more complex solver; using a base station for wind means adding extra transmissions and careful placement.

Table 5.1 below provides a summary comparison of these environmental compensation strategies, highlighting their strengths, weaknesses, key assumptions, and resource requirements.

Table 5.1 Comparison of environmental compensation strategies in acoustic positioning systems

Method	Strengths	Weaknesses	Assumptions
Sensor-based	Easy to implement; direct environmental readings	Extra hardware needed; limited spatial resolution	Sensors reflect actual path conditions
TOA estimation	No external sensors; adapts to changing conditions	Higher computation; needs more anchors	Sound speed is roughly uniform in space
Base station	Corrects directional wind errors; uses sound signals	Depends on wind uniformity; requires good placement	Wind is steady and uniform in the field

Another consideration is the assumptions about homogeneity in the environment. Each method assumes some level of uniformity. The real-time sound velocity estimation assumes the air within the area can be characterized by a single effective sound speed at any given time. This is quite reasonable in many indoor cases or small greenhouses. If the area is larger or has known temperature gradients, one might partition the space or use multiple simultaneous velocity estimates (e.g., one per region) to account for spatial variability. The base station wind compensation assumes the wind is uniform across the field. As evidenced by the two test setups (base station at center vs. at the perimeter), violating this assumption can reduce the benefit or even introduce new errors. Therefore, when deploying such methods, it is important to understand the variability of the environment. If multiple distinct microclimates or wind patterns exist, a single global compensation may not be sufficient, and a more distributed solution tailored to each sub-area may be required.

It's possible, and often desirable, to combine strategies. For instance, in an outdoor scenario it is possible to combine both algorithmic methods. A base station can be used to estimate wind, while the TOA algorithm can simultaneously estimate deviations in overall sound speed caused by temperature or humidity changes. In practice, one might handle temperature in a simpler way (e.g., by periodically measuring ambient temperature or leveraging predictable day-night cycles) and focus computational efforts on wind compensation, or vice versa. Important implementation considerations include ensuring you have enough anchors or beacons to support the algorithmic solutions, and that all nodes are properly synchronized or time-stamped for the time-of-flight measurements (see Chap. 3 on synchronization in acoustic systems). Using a common clock, or otherwise being able to measure relative timing between signals, is essential for multi-node methods like these.

In summary, compensation methods should be selected with regard to expected variability, required precision, and available resources. Simple sensor-based strategies may suffice for stable conditions, while algorithmic or hybrid approaches are needed when accuracy requirements are high or when temperature and wind fluctuate significantly. Designing for the worst-case conditions, validating assumptions with occasional ground-truth checks, and applying filtering to smooth transient effects further increase reliability. With careful planning, acoustic positioning systems can achieve centimeter-level accuracy even under challenging environmental conditions.

References

Tsay, L. W. J., Shiigi, T., Huang, Z., Zhao, X., Suzuki, T., Ogawa, Y., & Kondo, N. (2020). Temperature-compensated spread Spectrum sound-based local positioning system for greenhouse operations. *IoT, 1*(2), 2. https://doi.org/10.3390/iot1020010

Widodo, S., Shiigi, T., Than, N. M., Kikuchi, H., Yanagida, K., Nakatsuchi, Y., Ogawa, Y., & Kondo, N. (2014). Wind compensation for an open field spread spectrum sound-based positioning system using a base station configuration. *Engineering in Agriculture, Environment and Food, 7*(3), 3. https://doi.org/10.1016/j.eaef.2014.04.001

Chapter 6
Hybrid Systems and Sensor Fusion

Acoustic positioning systems or inertial measurement units (IMUs) could be used in indoor mobile robot localization. Acoustic systems, especially those based on SSSound, offer centimeter-level absolute positioning but suffer from motion-induced Doppler effects and time delays due to time-division scheduling. IMUs provide self-contained, high-rate motion data but are limited by cumulative drift over time. Hybrid systems that combine these modalities can mitigate each other's weaknesses, achieving both continuity and long-term accuracy. This chapter presents the foundational methods and architectures for such sensor fusion, with a focus on Kalman filtering and posterior probabilistic approaches. It also extends the discussion to multimodal systems that incorporate vision, LiDAR, and UWB ranging, highlighting their integration challenges and performance benefits in real-world environments such as greenhouses and factories.

6.1 IMU-Acoustic Sensor Fusion

6.1.1 Current Limitations of Single-Module Positioning Systems

Inertial navigation systems (INS) determine the position and orientation of a mobile platform by integrating raw motion data over time. The most common configuration is the strapdown IMU, which contains tri-axial accelerometers and gyroscopes. These sensors provide measurements of linear acceleration and angular velocity, respectively, relative to a local coordinate frame. The IMU's onboard processor (or the robot's computer) continuously integrates the acceleration to compute velocity, and integrates velocity to derive position. Similarly, angular rate measurements are integrated to yield orientation (attitude), i.e. roll, pitch, and yaw.

Z. Huang, *Acoustic and Signal-Based Local Positioning*, Navigation: Science and Technology 18, https://doi.org/10.1007/978-981-95-6083-7_6

While the measurement frequency of IMUs can exceed 100 Hz, making them ideal for capturing fast movements, their standalone use for positioning is fundamentally limited by drift. Sensor biases, scale factor errors, and environmental disturbances introduce cumulative errors that grow approximately quadratically with time if not corrected. For example, a small constant bias in the accelerometer will result in a position error that increases with the square of time. Thus, IMUs alone are not reliable for long-term localization, but they are extremely valuable for short-term motion estimation and for "gap filling" between absolute position updates from another system.

Acoustic positioning provides absolute position estimates with moderate temporal resolution (normally 1 Hz for TDMA and 4 Hz for FDMA), while IMUs deliver high-rate relative motion information but lack any long-term positional reference. Fusing the two allows the system to continuously track the robot's trajectory with both accuracy and continuity. In this hybrid arrangement, the acoustic subsystem periodically resets or bounds the drift of the inertial subsystem, and the inertial subsystem in turn interpolates and smooths the trajectory between acoustic. Table 6.1 summarizes the complementary characteristics of SSSLPS versus IMU-based dead-reckoning.

By leveraging this complementarity, a properly fused system can overcome the individual weaknesses of each modality. The acoustic signals act as global reference points to constrain the unbounded growth of inertial drift. Conversely, the IMU provides immediate responsiveness to motion and maintains the pose estimate during periods when acoustic updates are unavailable (e.g., if the robot is moving quickly or if acoustic signals are temporarily blocked). This complementary relationship is the theoretical cornerstone of IMU-acoustic fusion (Fig. 6.1).

To mitigate the near-far problem and to avoid mutual signal collision, anchors are often scheduled to transmit in separate time slots via a TDMA scheme. This approach prevents interference but introduces an inherent delay in acquiring a full set of range measurements, since the ToAs from all anchors are not available simultaneously. This temporal sparsity complicates localization under motion: the robot may have moved between successive anchor transmissions, leading to an inconsistency in the geometry for multilateration. Figure 6.2 illustrates a

Table 6.1 Comparison of acoustic and inertial measurement modalities

Feature	SSSLPS	INS
Measurement type	Absolute position (ToA-based)	Relative motion (acceleration/gyro)
Update rate	1–4 Hz	>100 Hz
Susceptibility	Affected by obstruction (line-of-sight needed)	Drift accumulates over time
Precision	High (centimeter-level)	High short-term resolution (low noise over short intervals)
Independence	Requires external infrastructure (beacons)	Self-contained (onboard sensors)

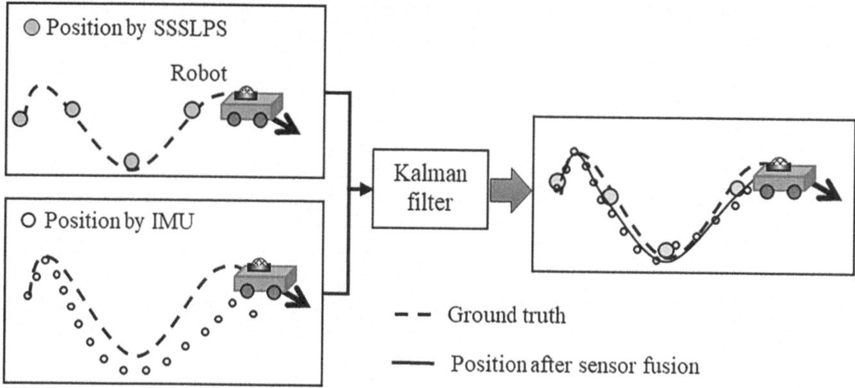

Fig. 6.1 Illustration of sensor fusion between SSSLPS and IMU

TDMA-based SSSLPS, where each anchor transmits in turn and the mobile robot registers sequential ToA measurements rather than simultaneous ranges. By the time the last anchor's signal arrives, the robot may have moved from the position it was at when the first signal arrived, introducing challenges for position calculation. As a result, the estimated position at the current time may span a larger uncertainty area than that of an FDMA-based system. The estimated region is marked in green in Fig. 6.2. The TDMA design thus trades reduced interference for increased update latency.

In practical terms, an IMU-acoustic fusion system can be thought of as an error correction loop: the IMU propagates the state with high frequency, and whenever an acoustic position fix comes in, the system corrects the estimated state to align with the absolute measurement. The result is a smooth and accurate trajectory. The performance gain can be significant. For example, Tientadakul et al. (2021) presented a soft-coupled fusion algorithm combining a TDMA-based ultrasonic local positioning system with an IMU for greenhouse robots, achieving a 100 Hz update rate. Their fusion approach substantially outperformed a conventional Extended Kalman Filter (EKF) in simulations and experiments, reducing horizontal and vertical position errors to 0.112 m and 0.116 m, respectively. This demonstrates that even with the asynchronous and slower updates of an acoustic system, clever fusion with inertials can yield accurate and responsive localization.

It is worth noting that similar sensor fusion strategies are widely used with other absolute positioning modalities as well. For instance, UWB radio localization combined with IMU is popular for indoor positioning; researchers have implemented such fusion using EKFs or particle filters, and found that the overall accuracy is largely limited by the accuracy of the absolute positioning system. In the following, we focus on the algorithmic frameworks for fusing acoustic (SSSLPS) and inertial data, which are generally applicable to other combinations like UWB + IMU or vision+IMU with appropriate modifications.

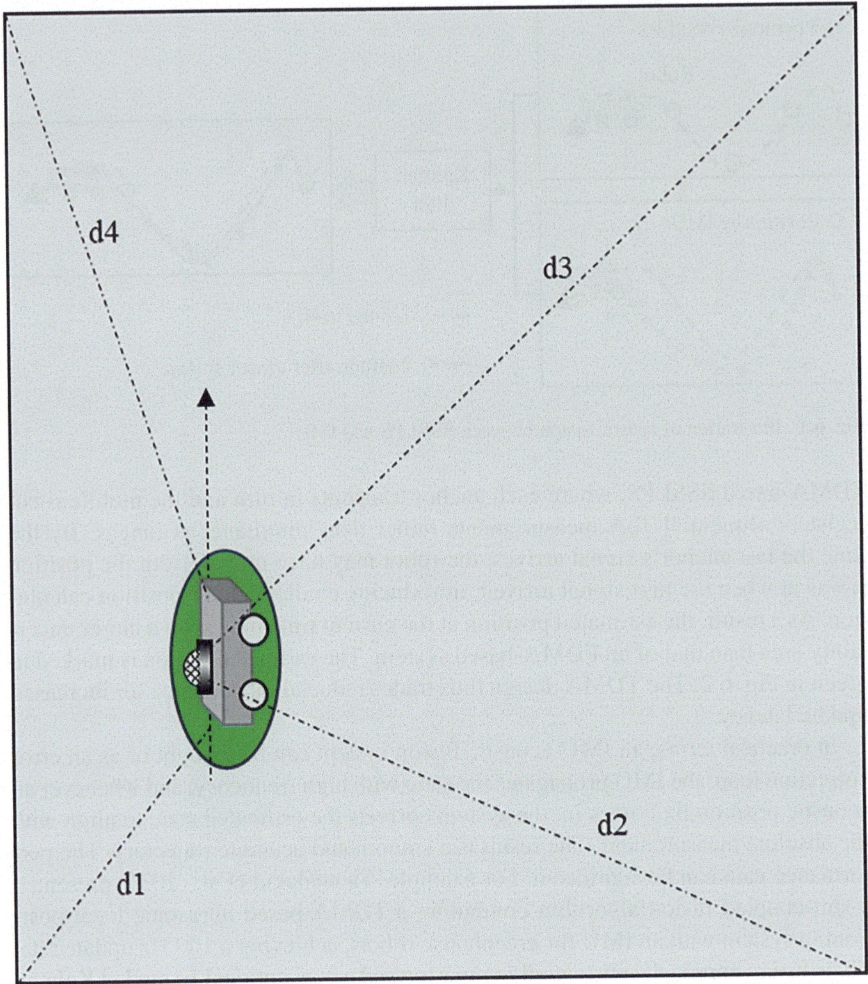

Fig. 6.2 Conceptual diagram of TDMA-based SSSLPS anchor transmissions and mobile robot ToA reception timeline

6.1.2 *Extended Kalman Filter–Based Fusion (FDMA Architecture)*

An EKF can integrate IMU dead-reckoning with spread-spectrum acoustic ranging measurements tightly. The SSSLPS with an FDMA scheme allows multiple anchor transmitters to emit distinct frequency-coded signals concurrently. Consequently, at each acoustic update epoch the robot's microphone can obtain ToA readings from all anchors simultaneously, enabling instantaneous multilateration. This high-level of synchrony lends itself to a classic EKF formulation wherein the state is corrected whenever a new batch of range measurements is available.

The EKF operates by alternating between prediction and update steps. The state vector typically includes the robot's kinematic variables (e.g. 3D position and velocity) and, if necessary, sensor bias terms (for example, accelerometer bias or an INS position offset). The filter prediction uses the IMU data to propagate the state forward in time, while the update uses acoustic range measurements to correct the state estimate. The process can be summarized as follows:

(a) **Prediction (Time Update)**

At each high-frequency IMU timestep (e.g. 100 Hz), the robot's state is predicted forward using the inertial measurements. Given the previous state x_{k-1} at time t_{k-1}, and control inputs or IMU readings u_{k-1} (accelerations a and angular rates ω, the system model $f(\bullet)$ computes a predicted state:

$$x_{k|k-1} = f\left(x_{k-1|k-1}, u_{k-1}\right) \tag{6.1}$$

This propagation integrates the accelerations to update velocity and position. The state transition also accounts for process noise (modeling uncertainties in dynamics and IMU errors). The error covariance is advanced accordingly. Notably, the acoustic sensor operates at a much lower rate (e.g. 4 Hz), so during many IMU cycles no new range observation is available. In those interim steps, the EKF relies purely on the IMU (INS) solution, and the state remains in prediction mode.

(b) **Measurement Update (Correction)**

Whenever the SSSLPS provides a new set of ToA measurements (typically from 4 anchors in our setup), the EKF performs a correction. The ToA readings are converted to range estimates z_i by multiplying by sound speed (with appropriate clock synchronization and environmental calibration). The measurement model $h(\bullet)$ predicts the expected ranges from the current predicted state $x_{k|k-1}$ to each anchor:

$$y_{p,k} = h\left(x_{k|k-1}\right) = \begin{bmatrix} d_1(x) \\ d_2(x) \\ \vdots \\ d_N(x) \end{bmatrix} \tag{6.2}$$

where, $d_i(x)$ is the distance from the robot's predicted position to the i_{th} anchor (for N anchors) and constitutes the predicted measurement vector $y_{p,k}$. Once the actual acoustic measurements $y_k = [z_1, z_2, \cdots, z_N]^T$ are received, the EKF computes the residual $r_k = y_k - y_{p,k}$ and applies the Kalman gain update:

$$x_{k|k} = x_{k|k-1} + K_k\left(y_k - y_{p,k}\right)$$
$$P_{k|k} = P_{k|k-1} - K_k H_k P_{k|k-1} \tag{6.3}$$

Here H_k is the measurement Jacobian (linearized matrix of partial derivatives of range w.r.t. state), and K_k is the Kalman gain optimized to minimize the state estimation error covariance. Intuitively, K_k weights the correction based on the relative uncertainty. If the inertial prediction is very uncertain relative to the acoustic measurements, the gain places heavy weight on the new acoustic data, and vice versa. The updated state $x_{k\,|\,k}$ thus blends the high-rate IMU-driven estimate with the absolute positioning information from SSSLPS. Importantly, if no new acoustic data arrived at time t_k, this update step is skipped (effectively $x_{k\,|\,k} = x_{k\,|\,k-1}$), and the filter continues propagating with the INS until the next measurement arrives.

The EKF fusion architecture combines high-rate inertial data and low-rate acoustic positioning to deliver robust and accurate localization for indoor robots (Fig. 6.3). The system begins with the IMU feeding high-frequency motion data, including accelerations and angular velocities, into an INS module, which continuously predicts the robot's position and velocity through integration. In parallel, a SSSLPS provides absolute range measurements, with each anchor fixed in the environment transmitting a unique spread-spectrum chirp signal.

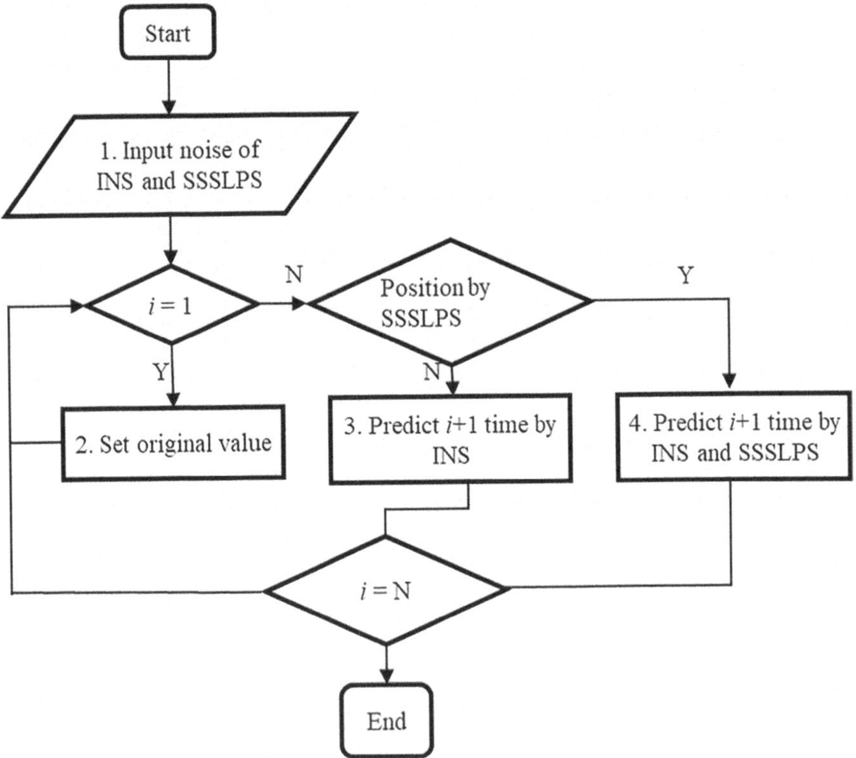

Fig. 6.3 Process diagram of EKF based sensor fusion

Under the FDMA scheme, all anchors transmit simultaneously on different frequencies, allowing the robot's microphone to detect and correlate multiple ToA signals in one update cycle.

These two data streams converge in the EKF estimator. The INS output drives continuous state prediction at the IMU rate, while acoustic range updates serve as asynchronous corrections. The EKF continuously loops through prediction steps, updating the state estimate with IMU data. When a new acoustic measurement frame is detected, the filter switches to correction mode, using the Kalman gain to optimally adjust the predicted state based on the residual between expected and observed ranges. If no acoustic data is available, the correction step is skipped, and the filter returns to prediction.

This tightly coupled design enables the system to produce high-rate pose estimates even when acoustic updates are intermittent. It accounts for sensor noise, timing misalignments, and acoustic signal delays by timestamping ToA inputs and aligning them with the correct prediction instance. Overall, this EKF-based fusion method leverages the stability and accuracy of acoustic localization with the continuity and responsiveness of inertial tracking, yielding a reliable navigation solution in complex indoor environments.

6.1.3 Posterior Probability Filtering Approach (TDMA Architecture)

In indoor positioning systems, combining the short-term accuracy of inertial navigation with the long-term stability of acoustic localization yields a powerful hybrid solution. This section introduces a tightly integrated method that fuses TDMA-based acoustic measurements from a SSSLPS with data from an INS. Unlike standard EKF approaches, which integrate sensor data at every time step, this architecture performs posterior fusion at discrete acoustic update moments, using the high-frequency INS output as prior state input.

The fusion system involves two major subsystems: the SSSLPS with TDMA protocol and a strapdown INS based on IMU data. Each acoustic anchor transmits in a dedicated time slot within a synchronized TDMA frame. The rover detects these spread-spectrum chirps and computes ToA measurements, which are then translated into range estimates. Concurrently, the IMU continuously outputs accelerometer and gyroscope data, which are integrated to estimate rover position and velocity through inertial dead reckoning. At each acoustic update (e.g., every 300 ms), the most recent INS-estimated pose serves as a prior, and a correction is applied based on the difference between expected and observed acoustic ranges.

Let $x_k = (x, y, z, v_x, v_y, v_z)^T$ denote the 6D rover state vector at time t_k, comprising position and velocity components. The INS provides this state at high frequency, while acoustic ToA measurements arrive intermittently. The fusion process includes the following steps:

(a) **INS Prediction**

At each IMU step t_i, the rover state is propagated forward:

$$x_{i+1} = f(x_i, u_i) + w_i \quad (6.4)$$

where u_i includes linear acceleration a_i and angular rate ω_i, and w_i is process noise. The prediction uses a standard strapdown integration:

$$v_{i+1} = v_i + a_i \Delta t$$

$$p_{i+1} = p_i + v_i \Delta t + \frac{1}{2} a_i \Delta t^2 \quad (6.5)$$

This prediction is repeated at every IMU update (e.g., 100 Hz).

(b) **Acoustic Range Observation**

At discrete times t_k, a full set of ToA-based range measurements $z_k = [d_1, d_2, \cdots, d_N]^T$ is received from N anchors. Each range d_j is computed as:

$$d_j = v \cdot \left(\tau_j^{recv} - \tau_j^{send} - \delta_t\right) \quad (6.6)$$

where v is the speed of sound, τ_j^{recv} is the reception time, τ_j^{recv} is the synchronized transmit time for anchor j, and δ_t is clock offset correction.

(c) **Measurement Model**

The expected range to each anchor given the predicted position $\hat{p}$ is:

$$\hat{d}_j = \left\| \hat{p} - p_j^{anchor} \right\| \quad (6.7)$$

The residual between observed and predicted ranges is:

$$r_k = z_k - h(\hat{x}_k) \quad (6.8)$$

where $h(\cdot)$ is the nonlinear measurement function mapping state to ranges.

(d) **Correction Step**

The posterior estimate is computed as:

$$x_{k|k} = x_{k|k+1} + K_k \left(z_k - h\left(x_{k|k+1}\right)\right) \quad (6.9)$$

The Kalman gain K_k is:

$$K_k = P_{k|k-1} H_k^T \left(H_k P_{k|k-1} H_k^T + R_k\right)^{-1} \quad (6.10)$$

where H_k is the Jacobian of the measurement function and R_k is the acoustic sensor noise covariance. The error covariance update is:

$$P_{k|k} = (I - K_k H_k) P_{k|k-1} \tag{6.11}$$

This update occurs only at ToA update times.

Accurate fusion relies on precise timing alignment. In TDMA-based SSSLPS, each anchor transmits at a pre-assigned time slot t_j^{slot}. The rover's clock is synchronized to the TDMA frame using GPS or network time protocol (NTP), and any residual drift is corrected by a software-controlled phase-lock loop. Each received chirp is matched to its slot index, ensuring that range estimation uses the correct transmit timestamp.

An important property of this setup is that all acoustic ranges are collected within a short time window $[t_k, t_k + \Delta T]$, making it suitable for batch update using EKF.

Figure 6.4 shows the conceptual architecture. The IMU continuously updates position and velocity via the strapdown INS. Periodically, a TDMA cycle completes, and multiple range measurements are received. These are fused with the INS-predicted pose via an EKF update. The posterior state is then passed back to the INS as a reset or bias correction reference.

The TDMA-based posterior fusion framework leverages the strengths of both acoustic and inertial systems. It uses the INS to bridge the gaps between acoustic updates and corrects drift using absolute ToA-based ranges. The Kalman filter formulation ensures optimal fusion under Gaussian noise assumptions, and the TDMA protocol provides scalable and collision-free range updates. This approach avoids overloading the filter with continuous updates and aligns well with practical deployment scenarios in robotic navigation and industrial indoor tracking.

This tightly coupled method is particularly suitable for settings where multiple robots share a common acoustic infrastructure, and where high-rate IMU data can be processed onboard while acoustic updates arrive over a shared wireless channel or network.

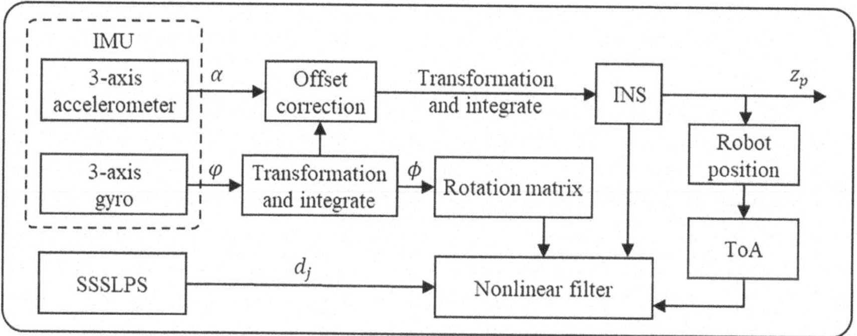

Fig. 6.4 Conceptual flowchart of the TDMA acoustic positioning and INS fusion algorithm

6.2 Dual Sensor Fusion

Another type of combination is to fuse SSSLPS with other local positioning systems, such as UWB radio, visual markers, or SLAM-based localization. In general, acoustic signals offer centimeter-scale ranging accuracy but require line-of-sight and quiet conditions, whereas RF and vision-based systems provide robustness in cluttered environments but may have coarser accuracy or drift over time. A dual-sensor fusion design uses algorithms (e.g. Kalman filters or graph optimization) to integrate these data streams into a single improved position estimate. For example, an EKF can update the robot's state x by blending an acoustic range measurement $z_k^{(acoustic)}$ with the prediction $\hat{x}$ based on another sensor (UWB, vision, or SLAM). The update takes the form:

$$\mathrm{x}_{k|k} = \mathrm{x}_{k|k+1} + \mathrm{K}_k \left(z_k^{(acoustic)} - h\left(\mathrm{x}_{k|k+1}\right)\right) \tag{6.12}$$

Through such sensor fusion techniques, hybrid systems achieve higher accuracy and reliability than any single sensor alone. Key architectures and methods for dual-sensor fusion are discussed below.

6.2.1 Fusion of SSSLPS and UWB Systems

Combining SSSLPS with an UWB ranging system yields a robust positioning solution that exploits the precision of acoustics and the all-weather operability of RF signals. UWB anchors (transceivers) installed around the target area, such as a greenhouse, can provide time-of-flight measurements even without visual line-of-sight, avoiding the GPS dropouts and occlusions caused by dense vegetation. SSSLPS can be integrated into this setup to further refine the accuracy. Acoustic signals (typically ultrasonic chirps) have shorter wavelengths and lower propagation speed than RF, enabling sub-decimeter ranging under ideal conditions. In fact, sound-based positioning in large indoor spaces like greenhouses has demonstrated superior accuracy to RF methods. However, ultrasound's range is limited, and it is easily blocked by obstacles or noise. A comparison of SSSLPS and UWB is shown in Table 6.2.

A logical fusion architecture is to use UWB for coarse localization and acoustic for fine positioning, capitalizing on each modality's strengths. The two data sources can be fused in a tightly-coupled filter or factor graph that accounts for their different error characteristics. For instance, a recently proposed factor graph optimization framework dynamically adjusts the weight of UWB measurements when multipath or NLOS conditions are detected, preventing UWB biases from corrupting the fused solution. In parallel, the acoustic ranging data (when available and reliable) get high weight due to their precision. Experiments in complex indoor environments show

Table 6.2 Comparison of SSSLPS and UWB

Feature	SSSLPS	UWB
Medium	Acoustic (sound)	Radio frequency (RF)
Accuracy	Very high (1–5 cm) in quiet LOS	Moderate (10–30 cm) even in NLOS
Line-of-sight (LOS)	Required	Not always required (penetrates walls)
Multipath sensitivity	High (echoes cause range errors)	Medium (RF reflections handled better)
Update rate	Low (limited by sound speed)	High (frequent RF pulses possible)
Noise susceptibility	High (affected by ambient sound)	Low (resistant to environmental noise)

that such multi-modal fusion can improve positioning accuracy by ~38% compared to a conventional Kalman filter using either sensor alone (Zhang et al., 2025). The fused system maintained ~12 cm RMSE even with only 40% of UWB packets received, thanks to acoustic constraints and IMU assistance suppressing drift.

Architecturally, acoustic–UWB fusion may require a common clock or synchronization strategy, since SSSLPS typically relies on precisely timed transmissions. One design uses an RF link to synchronize the emission of acoustic beacons with the robot's receiver clock. This kind of synchronization allows acoustic and UWB measurements to be time-aligned for integration. The fused data are often processed in an EKF or particle filter: UWB ranges provide global position updates, while acoustic ranges tighten the solution with fine detail. If the sensors update at different rates, the filter can incorporate slower acoustic readings as intermittent corrections to the higher-rate UWB or odometry estimates. Alternatively, a loosely-coupled approach might run separate estimators (one for UWB, one for acoustic) and then blend their outputs. However, a tightly-coupled fusion (directly combining raw measurements) generally yields better accuracy by using all information concurrently. The trade-off is increased system complexity—additional hardware and careful calibration are needed to synchronize clocks and coordinate frames between the RF and acoustic subsystems. Nevertheless, for safety-critical tasks or GPS-denied navigation conditions, this dual system provides a robust solution that works in diverse conditions (darkness or smoke won't impede ultrasound, while RF isn't affected by audible noise).

6.2.2 *Acoustic–Visual (Image Tag) Fusion*

Another effective dual-sensor strategy is fusing SSSLPS with a vision-based localization system, particularly those using fiducial image markers (e.g. AprilTags or ARToolKit tags) (Fig. 6.5). Visual tags placed at known positions in the greenhouse serve as artificial landmarks that a robot's camera can recognize to obtain an absolute position fix. Each tag encode its identity, and by detecting the tag in an image, the robot can compute the camera's 3D pose relative to that tag. This yields an

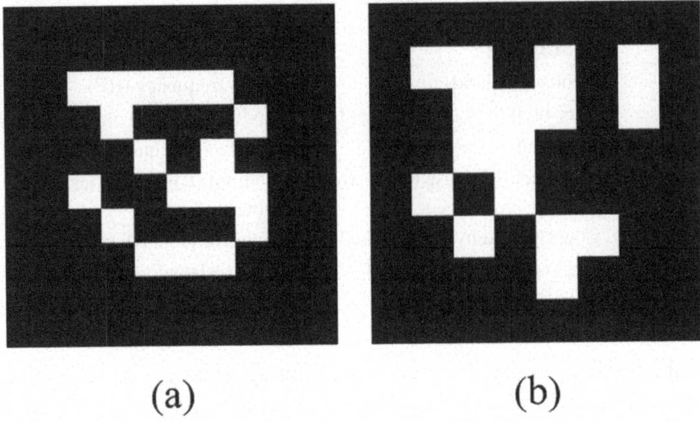

Fig. 6.5 Examples of the (a) ARTag and (b) AprilTag

external reference much like an anchor beacon, but using computer vision. The fusion of acoustic ranging with visual tag measurements combines two complementary sensing domains: acoustics are unaffected by lighting and can work over longer distances (tens of meters) even in featureless environments, while visual tags provide high local accuracy and orientation information when within camera view.

In an acoustic–visual fused system, the robot carries both a microphone/ultrasonic receiver and a camera. The greenhouse might be outfitted with a sparse grid of tags and acoustic emitters. Data fusion algorithms then combine the two sensing modalities. A simple approach is to use the visual tag detections to periodically reset or calibrate the acoustic system. For instance, if the robot's camera sees a tag and computes its absolute coordinates, the system can compare that to the SSSLPS-computed coordinates at the same moment and correct any drift or bias in the acoustic solution. Conversely, when the camera has no tag in sight (e.g. facing dense foliage or in low light), the ultrasonic ranging can continue to provide position updates. In practice, a Kalman filter or pose graph can treat tag observations as absolute position measurements with some uncertainty, similar to GPS updates, while treating acoustic ranges as continuous relative measurements. The fusion thus mitigates the weaknesses of each sensor. The acoustic LPS is immune to lighting conditions and provides continual coverage, but it can suffer from multipath or ambient noise, issues that do not affect vision. The camera can also help with orientation; a single acoustic measurement gives only distance, whereas a detected tag yields a full 6-DoF pose (position and heading) of the camera relative to the landmark. This information can improve the robot's heading accuracy when fused with, say, an IMU.

The benefits of acoustic–visual fusion are most pronounced in environments with intermittent visibility and intermittent acoustic clarity. For example, bright sunlight glare or fog might render tags hard to detect, but sound waves (especially ultrasonic) pass unaffected; conversely, a loud greenhouse fan or ultrasonic pest repeller could mask the SSSLPS beeps, but a camera would still function. By combining them, the robot can navigate through changing lighting and noise conditions. Moreover, visual tags are low-cost (printed markers) and can be added to existing structures without much infrastructure, complementing the more hardware-intensive acoustic beacons. Trade-offs include the need for additional processing (image analysis is computationally heavier and typically slower than ToF ranging) and the requirement to calibrate the camera and acoustic sensor coordinate frames. The system must know the precise location of each tag in the greenhouse coordinate system, a mapping effort that can be time-consuming and similar to surveying anchor locations. Maintenance is also a factor: tags can be obscured by dirt or plant growth and may need periodic replacement or cleaning. Despite these challenges, numerous robotics studies have validated that visual-inertial systems augmented with fixed landmarks achieve accurate, drift-free localization indoors. By extension, an acoustic–visual hybrid can ensure a mobile robot remains localized to within a few centimeters even after extended operation, using vision to eliminate long-term drift and acoustics to fill in gaps when the camera cannot observe landmarks.

6.2.3 Acoustic–SLAM Coupled Systems

A third category of dual-sensor fusion combines SSSLPS with an onboard SLAM system. SLAM algorithms (using LiDAR or camera data) allow a robot to localize itself relative to a map it builds of the environment, without relying on external infrastructure. For indoor robotic applications, 2D and 3D LiDAR SLAM have proven capable of achieving ~5–10 cm position accuracy for navigation by creating a floorplan of the rows and obstacles. However, SLAM-only solutions can suffer from cumulative drift (gradual position error growth) and even global misalignment. For instance, a robot might incorrectly close a loop in a greenhouse if rows are symmetrical (Fig. 6.6), causing a sudden jump in the estimated map. Integrating a few acoustic anchors into the SLAM framework provides absolute reference points that can pin the SLAM map to known coordinates and curb drift. Essentially, the acoustic system acts like an indoor GPS, periodically correcting the robot's position in the SLAM algorithm's internal map.

There are multiple ways to fuse acoustic data with SLAM. In a loosely-coupled approach, the robot could run SLAM independently to get a relative pose, and separately use SSSLPS to get an absolute position. These could then be blended to produce the final estimate. A more powerful method is tightly coupling acoustic ranges within the SLAM state estimation. Modern SLAM back-ends often use pose graph optimization or extended Kalman filters, which can incorporate additional sensors as extra constraints (Gong et al., 2024). In a graph-SLAM formulation, one can treat each acoustic beacon as a fixed node in the graph (with known global coordinates), and whenever the robot receives a distance measurement to that beacon, an edge (constraint) is added between the robot's pose node and the anchor node. This edge says, in effect, that the distance between the robot and anchor should equal the measured value (within some uncertainty). The graph optimization then tries to satisfy

Fig. 6.6 A typical greenhouse environment

these constraints along with the usual map feature constraints, resulting in a globally consistent solution. In EKF-SLAM, similarly, a range measurement can be handled by the measurement model $h(\mathrm{x})$ and used to update the estimated pose in the Kalman filter.

The fusion of SSSLPS with SLAM marries the reliability of global positioning with the flexibility of environment-driven localization. In a greenhouse, this means the robot can build a map of aisles, plants, and obstacles for path planning (a strength of SLAM) while never losing its absolute position within the global frame defined by acoustic beacons. Even if the robot traverses a long straight row with few distinctive features, the acoustic anchors ensure that its position within the overall greenhouse is maintained, thereby eliminating long-term drift. Moreover, if one or two acoustic anchors fail or go out of range, the SLAM odometry can carry the robot's pose estimation for a while, bridging the gap until another anchor comes into view. This redundancy enhances robustness. Additionally, fewer anchors might be required than in a stand-alone acoustic system; for example, instead of needing full 3D trilateration coverage everywhere, just a couple of well-placed acoustic beacons could constrain the SLAM map and prevent it from "floating."

6.3 Design Considerations and Recommendations

Deploying a multi-sensor positioning system in an indoor environment introduces several practical challenges that must be addressed in the design phase. The discussion is mainly focused on the greenhouse due to its unique environment with few environmental features for vision-based and SLAM-based positioning systems. Greenhouses are semi-structured, demanding environments: they contain metallic frames, rows of plants (which can obstruct signals), varying climate conditions, and ongoing agricultural operations. A hybrid SSSLPS + IMU or multi-sensor fusion system must be engineered for reliability under these conditions.

First, signal obstruction and multipath interference are significant challenges in greenhouses. Plants, structural supports, and equipment may partially or fully block acoustic signals, causing inaccurate measurements. Ultrasonic signals, in particular, are highly sensitive to obstacles and require clear line-of-sight. Similarly, UWB signals can penetrate foliage better but suffer from multipath interference due to reflections from metallic surfaces. Effective sensor fusion designs must, therefore, include sufficient redundancy and strategic placement of sensors to ensure consistent and reliable data collection despite these obstructions.

Second, temperature and humidity variations significantly impact acoustic positioning systems. The speed of sound in air varies with temperature (approximately 0.6 m/s per °C change), leading to errors if not properly compensated. In greenhouses, where temperatures fluctuate widely, accurate positioning requires real-time or frequent calibration of sound speed based on temperature and humidity sensors. Incorporating temperature sensors into the sensor fusion framework and dynamically adjusting the acoustic measurement model will significantly enhance positioning accuracy.

Third, greenhouse environments pose ambient acoustic noise from ventilation systems, irrigation pumps, and other machinery. Such noise elevates the baseline acoustic level, potentially degrading the signal-to-noise ratio of ultrasonic positioning signals. A robust spread-spectrum acoustic approach combined with sophisticated signal processing techniques (such as correlation-based filtering) can mitigate interference from ambient noise. Additionally, deploying sensors at optimal heights or orientations away from noise sources can further improve system performance.

Another critical consideration is system-level calibration and synchronization. Sensor fusion accuracy depends heavily on precise alignment between different sensor modalities. Misalignment between IMU and acoustic or UWB systems, for instance, can introduce significant positioning errors. Thus, initial and periodic recalibration of sensor positions, orientations, and time synchronization are essential. Methods such as automated calibration routines and external references (e.g., known physical markers or robotic calibration paths) help maintain high accuracy throughout operation.

Lastly, robust greenhouse robotics require careful planning for maintenance and reliability. Sensor fusion components in greenhouse environments must withstand moisture, dirt, chemical exposure, and physical impact from daily agricultural operations. Choosing ruggedized sensor enclosures, weatherproof connectors, and durable mounting solutions will minimize maintenance. Periodic inspections, recalibration procedures, and straightforward replacement processes for damaged sensors further ensure sustained system accuracy and reliability.

Table 6.3 Comparison between typical indoor positioning systems

Feature	SSSLPS (acoustic)	UWB	Visual/ SLAM-based	Why fusion helps
Accuracy	High (1–5 cm)	Moderate (10–30 cm)	High (feature-dependent)	Blends precision with resilience
Line-of-sight (LOS)	Required	Not strictly needed	Required for markers/features	UWB covers NLOS scenarios, visual supplements orientation
Multipath sensitivity	High	Moderate	Low	Cross-verification reduces error
Update rate	Low	High	Moderate to high	UWB and visual data fill acoustic gaps
Noise susceptibility	High (ambient acoustics)	Low	Moderate (lighting dependent)	Mitigates interference and noise issues
Environmental impact	Temperature sensitive	Less sensitive	Light and visibility sensitive	Compensation through cross-calibration

Table 6.3 summarizes these sensor characteristics clearly, highlighting the practical necessity of sensor fusion based on various environmental and operational conditions typical in greenhouses.

In conclusion, hybrid positioning systems in greenhouse robotics should address signal obstruction and multipath interference, sensor synchronization, environmental compensation, and ease of maintenance. The careful integration of complementary positioning technologies through sensor fusion addresses these challenges effectively, delivering reliable, precise localization performance critical to autonomous greenhouse robotics.

References

Gong, L., Gao, B., Sun, Y., Zhang, W., Lin, G., Zhang, Z., Li, Y., & Liu, C. (2024). preciseSLAM: Robust, real-time, LiDAR–inertial–ultrasonic tightly-coupled SLAM with ultraprecise positioning for plant factories. *IEEE Transactions on Industrial Informatics, 20*(6), 8818–8827. https://doi.org/10.1109/TII.2024.3361092

Tientadakul, R., Nakanishi, H., Shiigi, T., Huang, Z., Tsay, L. W. J., & Kondo, N. (2021). Spread Spectrum sound with TDMA and INS hybrid navigation system for indoor environment. *Journal of Robotics and Mechatronics, 33*(6), 6. https://doi.org/10.20965/jrm.2021.p1315

Zhang, F., Li, J., Zhang, X., Duan, S., & Yang, S.-H. (2025). Indoor fusion positioning based on "IMU-ultrasonic-UWB" and factor graph optimization method (version 1). arXiv. https://doi.org/10.48550/ARXIV.2503.12726.

Chapter 7
Implementation and Experimental Validation

This chapter presents the implementation and experimental validation of the SSSLPS in both wired and wireless configurations. It describes the experimental setups used to evaluate system performance in both indoor laboratory and greenhouse environments, explains the procedures for obtaining reliable ground truth data, and outlines the methodology for assessing the accuracy and robustness of the SSSLPS. Key aspects such as geometry-induced error amplification, motion-induced Doppler effects, and signal processing evaluation are addressed through a series of carefully designed tests and quantitative analysis.

7.1 Experimental Considerations and Setup

This section details the integration of the SSSLPS in both wired and wireless configurations, and describes how the system was set up in real-world environments. We discuss the hardware components and signal flow for each configuration, the spatial deployment of speakers and microphones, and the sound signal parameters (carrier frequencies, modulation scheme, chip rates, and multiple-access scheduling) used in our experiments. Key environmental challenges such as multipath sound reflections and temperature-driven sound velocity variations are also highlighted, along with the measures taken to address them.

7.1.1 Wired SSSound Localization System

In the wired implementation of the SSSLPS, all acoustic hardware is connected via cables to a central processing unit for tight synchronization and control. Figure 7.1 illustrates the architecture of the wired system. A central PC generates both the

Z. Huang, *Acoustic and Signal-Based Local Positioning*, Navigation: Science and Technology 18, https://doi.org/10.1007/978-981-95-6083-7_7

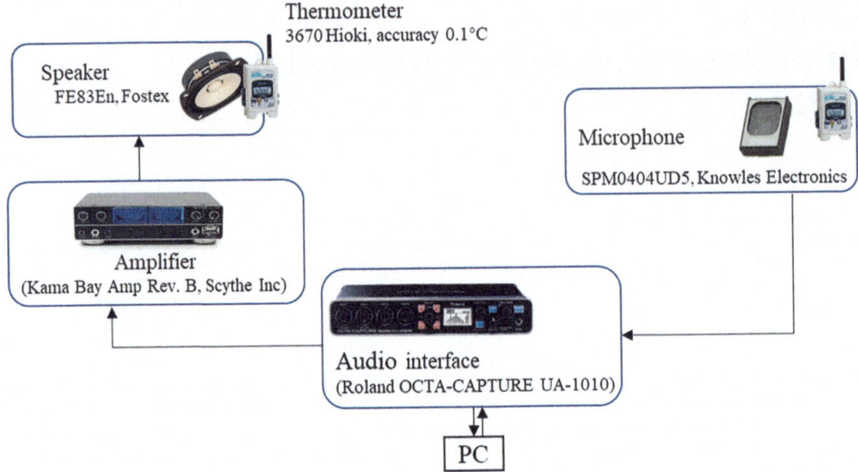

Fig. 7.1 Schematic of the wired SSSound positioning system

baseband spread-spectrum acoustic signals and a trigger pulse. These signals are output through a multi-channel audio interface (for example, a Roland OCTA-CAPTURE UA-1010), which performs digital-to-analog conversion (DAC) at a high sampling rate (typically 96 kHz, 16-bit) to cover the ultrasound frequency band used by the SSSound signals. In the typical system, for instance, a 24 kHz carrier modulated by a 12 kcps M-sequence produces an ultrasonic spread-spectrum signal spanning approximately 12–36 kHz. Four power amplifiers (one per channel) boost the analog outputs and drive four high-frequency speakers (tweeter transducers) that serve as fixed acoustic beacons. Concurrently, a synchronization trigger signal (a short electrical pulse on a dedicated channel) is sent via cable to the receiver subsystem so that transmit and receive timelines are precisely aligned.

The choice of acoustic transmitters (speakers) significantly influences the system's range, bandwidth, and signal fidelity. Common options include piezoelectric ultrasonic transducers (typically resonant around 40 kHz) and conventional audio tweeter speakers (which operate in the audible to ultrasonic range). Piezo ultrasonic transducers (such as the Murata MA40 series, ~10 mm in diameter) are inexpensive and have a narrowband response centered near 40 kHz. They can emit high SPLs around their resonant frequency (e.g. ~120 dB SPL at 40 kHz) with a moderate beam width of about 60–80°, which is useful for focused coverage. However, their narrow bandwidth is a limitation: these transducers tend to "ring" at the resonant frequency and cannot easily transmit wideband signals or frequency-hopping sequences. In practice, this means advanced modulation techniques like FHSS are challenging or costly to implement with basic ultrasonic buzzers. Audio tweeters, on the other hand, can cover a broad frequency range. This allows the transmission of wideband audible chirps or coded signals, providing broader bandwidth. The drawback of using audible frequencies is the potential intrusiveness (the sound may be heard by humans and is susceptible to ambient noise). A useful compromise is to use high-frequency tweeters or ultrasonic emitters that operate just above the human hearing range (e.g. 20–40 kHz). These can generate inaudible spread-spectrum signals while still offering greater bandwidth than a resonant piezo transducer. For instance, some indoor positioning systems repurpose off-the-shelf speakers or even smartphone speakers to emit coded chirps in the 18–22 kHz range, leveraging readily available hardware at the cost of limited range.

On the receiver side, the microphones can likewise be either specialized ultrasonic sensors or general broadband acoustic sensors. Standard electret or Micro-Electro-Mechanical Systems (MEMS) microphones often have an upper frequency limit around 18–20 kHz (suitable for audible sound). However, extended-range versions exist. For example, certain MEMS microphones such as the Knowles Electronics SPU0410LR5H-QB-7 have a usable response up to approximately 40 kHz. Many indoor ultrasound positioning systems use dedicated ultrasonic receiver modules. A common design pairs a 40 kHz piezoelectric transducer with a high-gain amplifier circuit to provide basic signal conditioning and detection. Such modules are very cost-effective for short-range ranging and positioning tasks, though they may sacrifice bandwidth. In our wired SSSLPS, we use high-bandwidth microphones capable of sensing the 12–36 kHz band of our signals (Huang et al., 2017), ensuring that the spread-spectrum codes can be received with fidelity.

To synchronize measurements in the wired system, the PC outputs a trigger signal along with the acoustic signals. Specifically, channels 1–4 of the audio interface carry the spread-spectrum sound signals for the four speakers, and channel 5 carries only the trigger pulse (as depicted in Fig. 7.2). In the wired configuration, this trigger (channel 5) is physically looped back to an input of the audio interface that connects to the receiver's processing unit. The microphones' data acquisition is started

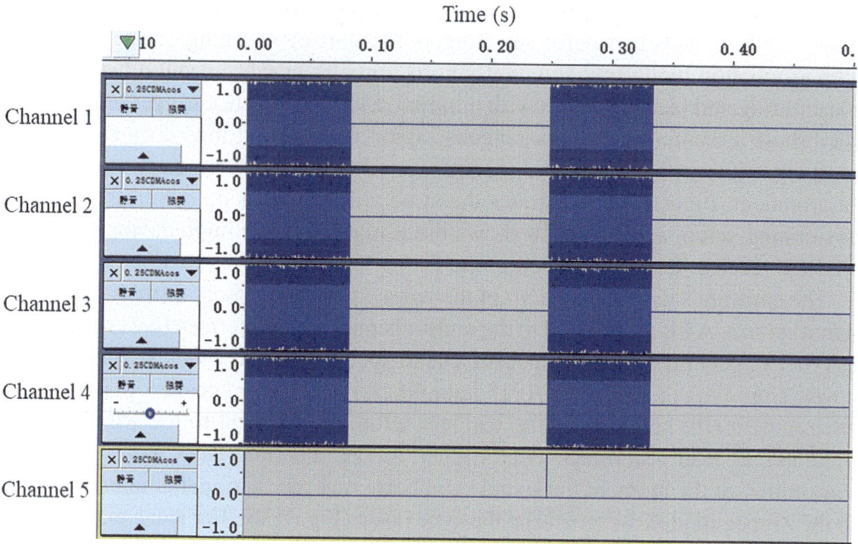

Fig. 7.2 The five channels of emitted SSSound

slightly before the sound emission, and when the trigger pulse is detected in the recording, it serves as a timing reference for time-of-flight measurements. Because the trigger travels over a wire with negligible delay, transmit and receive events can be aligned with virtually no uncertainty in timing. In other words, the wired trigger ensures there is effectively zero time synchronization error between the emission of the sound signal and the start of reception at the microphone. This hardware-triggered synchronization is critical for accurate TOA calculation, as it removes the need to estimate or correct for any trigger latency.

7.1.2 Wireless SSSound Localization System

A wireless version of the SSSound positioning system was also developed to increase the system's mobility and ease of deployment. In the wireless configuration, the fixed infrastructure (speakers) still emits the same acoustic signals as in the wired setup, but the receiver hardware on the moving target operates untethered. Synchronization of timing between transmit and receive units is achieved not through a physical cable, but via a wireless trigger signal. Essentially, the system is split into two units: (1) a sound-emitting unit at the fixed base station (including the central PC and speakers), and (2) a battery-powered wireless receiver unit mounted on the robot or vehicle to be tracked.

In a wireless SSSLPS, the trigger signal is conveyed by a low-latency radio frequency link. The Zigbee transceivers (3S System 3sZ11 modules operating in the 2.4 GHz band) is employed for this purpose. The base station's Zigbee transmitter emits a short wireless trigger pulse contemporaneously with the start of acoustic signal playback. On the robot's receiver unit, the corresponding Zigbee module immediately detects this pulse and triggers the start of sampling on the receiver's data acquisition hardware. This mechanism provides synchronization between the transmitting and receiving ends with minimal delay. The Zigbee trigger broadcast is very short (essentially an instantaneous digital packet) and thus does not significantly interfere with the acoustic spectrum or other wireless communications in the environment. Figure 7.3a illustrates the system diagram for the wireless SSSound positioning setup, and Fig. 7.3b shows the actual hardware implementation of the wireless receiver and base station units.

The emitting side (base station) of the wireless system closely mirrors that of the wired system. A PC, connected to the multi-channel audio interface (e.g., the Roland UA-1010), generates four channels of spread-spectrum audio which are fed through power amplifiers (e.g., Fostex AP15d amplifiers) to the four speakers placed in the environment (the speakers in the wireless setup are identical to the wired setup's beacons). Instead of a direct wired trigger, the PC also interfaces with the Zigbee transmitter; at the moment the sound signals are sent out, a trigger command is sent to the Zigbee module to broadcast the sync pulse (Fig. 7.3b). For some experiments (especially indoor ones), the base station PC was additionally connected to an external motion capture system (described later) via a data link, enabling simultaneous recording of ground-truth position data at 100 Hz for evaluation.

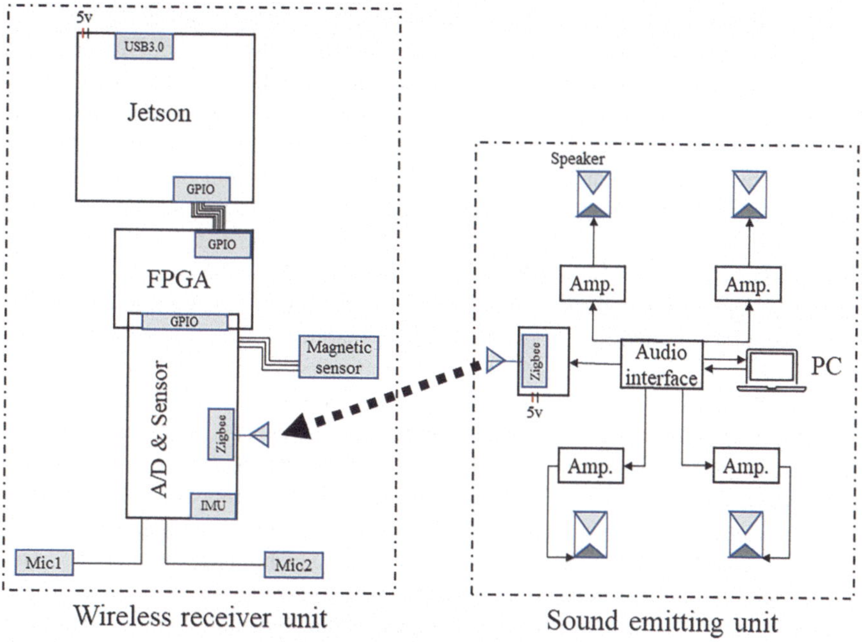

5v
USB3.0
Jetson
GPIO
GPIO
FPGA
GPIO
Magnetic sensor
A/D & Sensor
Zigbee
IMU
Mic1
Mic2
Wireless receiver unit
Speaker
Amp.
Amp.
Zigbee
5v
Audio interface
PC
Amp.
Amp.
Sound emitting unit

(a)

Jetson
FPGA
Mic1
Zigbee
Mic2
Magnetic sensor
Zigbee

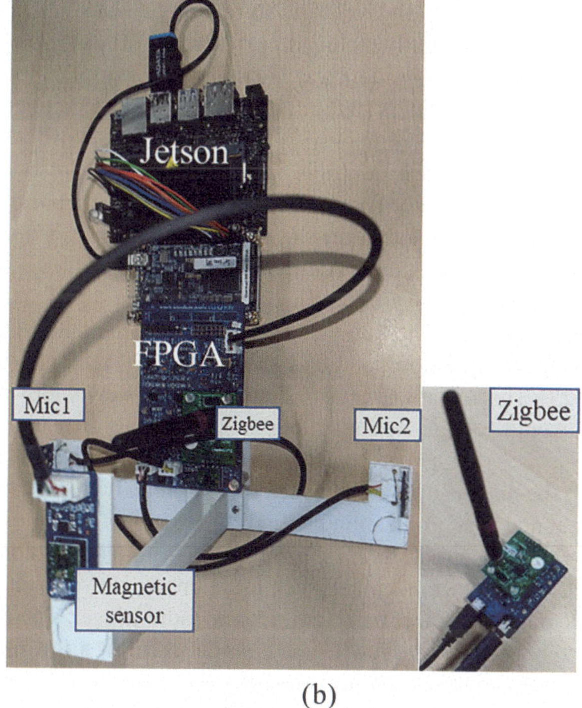

(b)

Fig. 7.3 (**a**) Diagram example and (**b**) hardwires of the wireless SSSound positioning system

The wireless receiver unit mounted on the robot is a compact integration of sensors, processing, and wireless communication. In our implementation, this receiver unit consists of a custom ADC and sensor board mated with an FPGA board and an NVIDIA Jetson embedded computer for processing. The sensor board includes two MEMS microphones (Knowles SPU0410LR5H-QB omni-directional microphones) separated by a fixed distance on the robot's frame, an inertial measurement unit (IMU, InvenSense MPU-9250), a 3-axis magnetometer (Honeywell HMC5883L), a barometric pressure sensor (TE Connectivity MS5611), and a digital thermometer (Analog Devices ADT7410). The dual microphones allow the system to receive ultrasonic signals at two distinct points on the robot, which can be exploited to estimate orientation in addition to position. This capability has been demonstrated in prior work (Huang et al., 2020), although orientation estimation is beyond the scope of this chapter. All sensor signals feed into a Cyclone IV FPGA (Altera EP4CE22F17C6N) on the board via general-purpose I/O. The FPGA handles real-time data acquisition (performing analog-to-digital conversion for the microphones) and can execute preliminary signal processing such as buffering and possibly hardware correlation. The digitized acoustic data (and other sensor readings) are passed to the on-board NVIDIA Jetson computer through a high-speed interface. The Jetson runs the main localization algorithm: it computes cross-correlations of the received signals against locally stored copies of the transmitted spread-spectrum codes to determine the TOA from each speaker (similar to how the PC did in the wired system). With at least three TOAs (one from each speaker's signal), the Jetson then solves the spatial localization problem to obtain the robot's 3D position (and, if using the two mics, potentially the heading). Because all processing is done on-board the robot, continuous communication with the base station is not required; the only wireless exchange needed is the initial trigger synchronization at the start of each measurement cycle.

One challenge in the wireless design is balancing on-board processing load with the system's desired update rate. The SSSLPS is intended to work over ranges up to roughly 50 m. The one-way sound propagation time for 50 m is about 0.15 s (at ~340 m/s sound speed). If we budget additional time for signal processing and computation (say ~0.1 s per cycle), a full cycle of capturing the signal and computing a range from one speaker could be on the order of 0.25 s. In an FDMA system or a fast TDMA cycle, the position solution can be obtained at approximately 4 Hz (Tsay et al., 2023). This update rate is sufficient for many slowly moving robots, but higher speeds would require optimizations in processing or a higher degree of parallelism.

In summary, the main difference between the wired and wireless systems lies in their deployment and synchronization method. The wireless system is more practical for real-world applications, as it frees the robot from tethered connections and allows more flexibility in movement over large areas. However, it introduces a potential source of timing uncertainty, namely the radio trigger. The Zigbee trigger was observed to have generally low latency, on the order of a few milliseconds or less, although slight variations in timing were present. Therefore, before each experimental deployment of the wireless system, we calibrated the Zigbee trigger timing to ensure the receiver's sampling starts at the correct offset relative to the sound emission. In contrast, the wired system, with its direct cable synchronization, has virtually no timing jitter and thus served as a reliable platform for foundational research and baseline performance tests. In scenarios where ultimate accuracy and synchronization certainty are needed (e.g., controlled lab experiments focusing on algorithm development), the wired setup is preferable. For field deployments and mobile applications, the wireless setup is far more convenient despite the need for an initial sync calibration.

7.1.3 Deployment in Indoor and Greenhouse Environments

The SSSLPS has been tested in two types of environments, namely an indoor laboratory setting and a greenhouse, each of which presents unique challenges for acoustic positioning. Several prior studies on indoor positioning guided the deployment strategy in these environments. Here the practical considerations and setup configurations are outlined for both scenarios.

When deploying the system in a greenhouse environment, the primary challenges are multipath effects and NLOS propagation caused by the plants and structural elements. To mitigate these issues, we positioned the speakers and microphones to maximize line-of-sight paths. One effective approach was to elevate the speakers and receiver above the height of the plants. By placing the acoustic beacons overhead (for example, suspending them from the greenhouse ceiling or mounting on tall stands) a clear direct path to the robot on the greenhouse floor can be created, similar to how UWB positioning systems have been deployed in greenhouses (De Preter et al., 2018). In cases where the robot operates on a bench-top cultivation system (raised platforms), the sensors can also be placed slightly above the ground (e.g. ~10 cm above the surface). This height is small compared with the acoustic wavelength. For ultrasound at 24 kHz, the wavelength is approximately 14 mm in air, so such placement effectively allows the sound to hop over small obstacles. This approach is analogous to certain UWB deployments that are designed to avoid crop obstructions (De Preter et al., 2018). The goal in the greenhouse setup is to maintain as clear a line-of-sight as possible between each transmitter (speaker) and the receiver, thereby reducing multipath interference.

Another consideration in the greenhouse is the SPL required. The greenhouse is a relatively large, open space (often tens of meters in each dimension), and background noise levels can vary (e.g., due to ventilation fans, rustling plants, machinery). The speakers' maximum output is around 110 dB SPL measured at 10 cm from the speaker. Such a level in practice allowed the signals to be detectable at distances of 50–60 m or more in our tests (the exact effective range depends on the modulation and the ambient noise). In the greenhouse deployment, we typically drove the speakers at or near their maximum SPL to ensure sufficient range and signal-to-noise ratio, given that we wanted to cover an area on the order of 20–30 m across for robot operations. We did not observe issues with the plants from the ultrasound, as the frequencies used were inaudible to humans and likely harmless to plants (the energy levels were low and spread-spectrum signals minimize peak power).

A further environmental factor in greenhouses involves the variation in sound speed due to temperature differences and air currents. Climate control systems commonly found in greenhouses, such as heating, cooling, and ventilation fans, can introduce spatial temperature gradients as well as steady airflow. These factors influence the propagation speed of sound, which increases with rising temperature and can be slightly altered by wind through advection. To account for temperature-related variations, a thermometer integrated into the wireless receiver's sensor suite was used to measure the ambient temperature, enabling the assumed sound speed to be adjusted accordingly for accurate distance estimation.

Wind effects present a greater challenge. During testing in greenhouse environments, a constant and gentle airflow was typically observed due to the operation of ventilation systems. To mitigate the potential impact of wind on acoustic positioning, the coordinate system was oriented so that the primary wind direction aligned with one axis, and measurements were repeated in both directions along that axis. This allowed any directional bias introduced by the wind to be statistically averaged out. For applications requiring higher precision, the integration of additional instruments such as anemometers or spatially distributed microphones may be employed to estimate and correct for wind-induced effects. It should be noted that even modest and steady airflows—on the order of a few meters per second—can introduce non-negligible bias in position estimation and should be considered in high-accuracy scenarios.

Indoor laboratory environments present a distinct set of challenges compared to greenhouse settings. Rigid boundaries such as walls, ceilings, and floors generate pronounced echoes and facilitate multipath signal propagation. Multipath effects in indoor conditions have been observed to be more severe than in greenhouses (Huang et al., 2021), where the larger volume and the presence of sound-absorbing vegetation result in more attenuated reflections.

To ensure accurate localization under such conditions, signal processing techniques capable of distinguishing the direct-path signal from overlapping reflections must be employed. High-resolution correlation processing has been utilized to enhance discrimination of arrival times. Additionally, speaker placement was strategically designed to reduce the likelihood of symmetric echo paths. This included varying the heights of speakers and orienting them away from direct-facing wall alignments, thereby avoiding scenarios in which reflected signals arrive at the receiver simultaneously with the direct path. In post-processing, temporal windowing was applied to exclude correlation peaks that occurred beyond the anticipated arrival window, effectively suppressing long-delay echoes and improving positioning reliability.

Another factor relevant to dynamic indoor experiments is the Doppler effect resulting from receiver motion. As the receiver-equipped robot moves, a frequency shift occurs in the received sound signal (Doppler shift), which can distort the correlation peak used for TOA estimation. To mitigate this distortion, a DSC algorithm, described later in Sect. 4.2.3, was employed to dynamically adjust the reference signal frequency in accordance with the observed Doppler shift. This approach was particularly essential in scenarios involving high carrier frequencies or rapid robot motion, where even moderate velocities produce significant frequency deviations.

For indoor deployment, speaker output levels and signal designs were adjusted to account for the presence of human occupants. Given the confined operating range (approximately 10–20 m), maximum sound pressure output (up to 110 dB SPL) was deemed unnecessary. Instead, speaker levels were set to approximately 70–80 dB SPL at 10 cm, which was sufficient to achieve reliable positioning within the required range while reducing audible disturbances. To further minimize perceptibility, signal frequencies above approximately 15 kHz were utilized. Because human auditory sensitivity declines markedly above this threshold, the use of ultrasonic carriers (e.g., 24 kHz) and the suppression of strong audible spectral components enabled the system to function in occupied indoor spaces without generating discomfort.

In conclusion, indoor deployments necessitated rigorous handling of multipath interference and Doppler-induced distortions, due to confined environments and dynamic movements. By contrast, greenhouse deployments placed greater emphasis on maintaining unobstructed acoustic paths and compensating for environmental variability such as temperature gradients and airflow. These strategies collectively supported reliable SSSLPS operation across both experimental contexts, as discussed in the following sections.

7.2 Ground Truth Estimation Methods

To rigorously validate the positioning accuracy of the SSSLPS, position estimates must be compared against a reliable ground truth reference. In GPS-denied environments such as indoor or enclosed spaces, several alternative positioning systems are capable of providing this reference. In the present validation process, three primary types of reference systems can be employed: (1) a robotic total station, (2) an optical motion capture system, and (3) a posture sampling sheet. Each system is characterized by distinct operating principles and performance parameters. To ensure suitability as a benchmark, the selected reference system must exhibit both higher accuracy and sufficient temporal resolution compared to the system under evaluation. The procedures for synchronizing data from the reference systems with SSSLPS outputs, as well as the methods used for coordinate transformation into a unified spatial frame, are detailed in the following sections.

7.2.1 Ground Truth Estimation Techniques

A total station is a laser-based surveying instrument capable of highly precise distance and angle measurements. It can be utilized to track a prism target mounted on a moving robot. Modern total stations support measurement ranges up to approximately 1500 m, with typical accuracy on the order of ±

(1.5 mm + 2 ppm × distance). For instance, a 50 m distance results in an error of approximately 1.6 mm. A Sokkia total station was deployed in the greenhouse to obtain ground truth positions. As shown in Fig. 7.4a, it presents the total station instrument, and Fig. 7.4b displays the 360° reflective prism mounted on the robot. The total station continuously tracks the prism and measures the range, or line-of-sight distance, together with the Zenith Angle (ZA), which is the vertical angle from the station with 0° indicating straight up and 90° indicating horizontal, and the Horizontal Angle Right (HAR), which represents the azimuth angle measured clockwise from a defined reference direction. The instrument's onboard computer or an external data collector converts these polar coordinates into Cartesian coordinates (X, Y, Z) relative to the total station's position. Figure 7.4c illustrates an example display readout with local X, Y, Z coordinates, slope distance, and angular values for the prism. During the experiments, the total station provided position

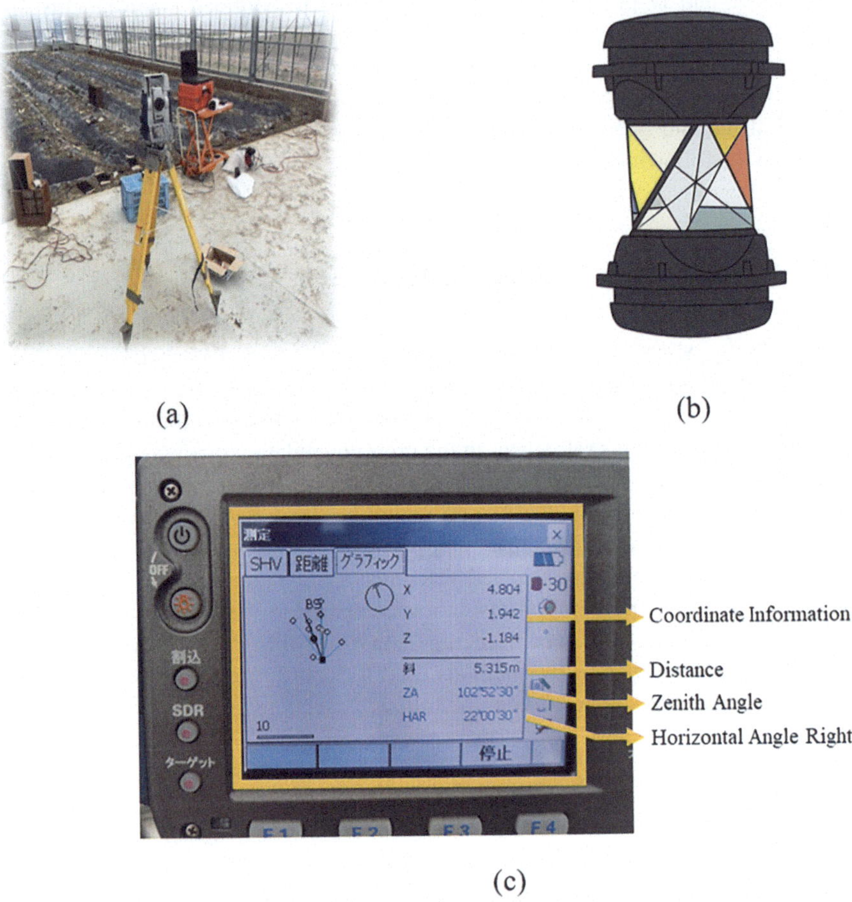

Fig. 7.4 (**a**) A total station set up in a greenhouse; (**b**) a 360° reflective prism target mounted on the robot, and (**c**) display of the measurement

updates at approximately 5–10 Hz, and the data were logged via Bluetooth to a laptop for synchronization with the SSSLPS.

While offering excellent accuracy for static or slow-moving targets in open areas, the total station exhibits some limitations. The update rate remains relatively low (typically below 10 Hz), which may be inadequate for capturing fast dynamic motion. Additionally, continuous operation requires an unobstructed line-of-sight to the prism. Any occlusion caused by greenhouse posts, vegetation, or structural elements interrupts tracking. As a result, experiments required the robot to operate within clear paths to maintain visibility. The total station is therefore most appropriate for static calibration or dynamic tests in minimally obstructed environments.

Optical motion capture systems serve as a suitable alternative for indoor experiments involving rapid robot movement or requiring high-frequency tracking. A Vicon motion capture system (Vicon Industries Inc.) consists of multiple infrared Time-of-Flight cameras deployed throughout the laboratory. These cameras track the 3D positions of retro-reflective markers affixed to the robot. The system achieves sub-millimeter accuracy (approximately 0.1–1 mm) and provides high temporal resolution, with a typical update rate of 100 Hz. Figure 7.5 displays the Vicon software interface visualizing real-time marker tracking. The marker cluster mounted on the robot is identified as a rigid body, and the robot's pose is streamed to the processing computer. These systems are widely used in robotics for ground truth evaluation, particularly in controlled laboratory settings. Limitations include sensitivity to infrared interference, which restricts operation in direct sunlight, as well as a limited tracking volume determined by the number and placement of cameras. In typical deployments, the usable capture area spans approximately 5 m × 5 m, enabling free movement of the robot within that region.

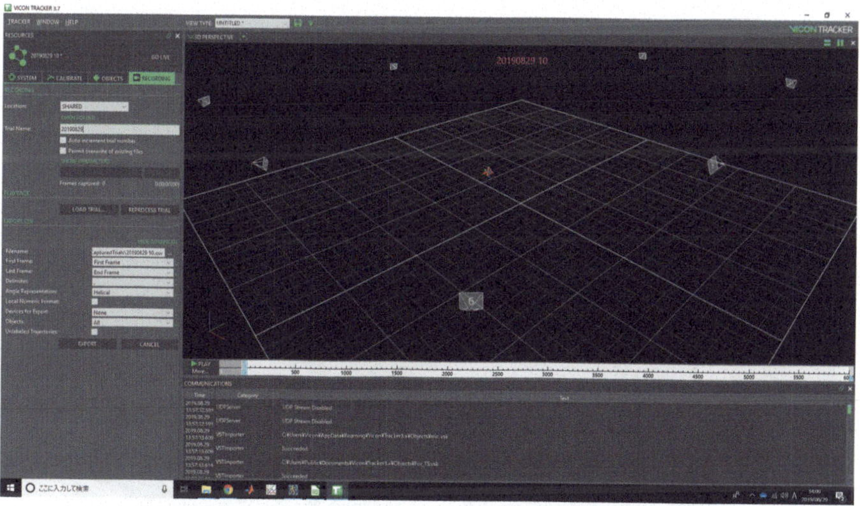

Fig. 7.5 Screenshot of the Vicon motion capture system interface

Accurate tracking is facilitated by placing markers on the robot in a defined configuration (Fig. 7.6). Four markers are arranged as follows: markers (1) and (2) are spaced horizontally on the top surface of the robot to indicate heading; marker (3) is located near the IMU module to support orientation comparisons; and marker (4) is elevated to ensure consistent visibility and aid in unique identification. This configuration is used by the Vicon software to define a rigid body and enables precise position and orientation tracking. Reliable results require secure marker attachment and proper calibration. Once configured, the system provides consistent and high-fidelity ground truth data.

In terms of cost and practicality, total station and motion capture systems span opposite ends of the spectrum. Total stations are relatively portable and available at moderate cost (several thousand USD), while professional motion capture systems incur significantly higher costs and require a dedicated indoor setup. Accordingly, the total station is applied in outdoor or greenhouse environments, whereas the Vicon system is reserved for indoor laboratory experiments.

A posture sampling sheet, introduced as a cost-effective alternative, is also applied for discrete validation (Jiang et al., 2022). This method involves a floor mat marked with a known grid pattern. When the robot halts on the mat, its position is determined based on labeled positions (either visually or via embedded sensors). While this approach does not support continuous tracking, it enables evaluation of steady-state accuracy at specific checkpoints. Planned robot trajectories incorporate stops at known positions, allowing comparison between mat labels and SSSLPS estimates. Originally designed to evaluate navigation and control performance, this method provides useful coarse accuracy validation. Observed errors at checkpoint positions typically fall within a few centimeters.

Fig. 7.6 Example of marker placement on the robot for motion capture

Additional alternative reference system such as SLAM-based odometry tools (e.g., EVO by Yan et al., 2022) and UWB localization systems are also under consideration. The EVO tool fuses wheel odometry with visual or LIDAR-based SLAM to generate trajectory estimates with approximately 1.5 cm accuracy. While not directly used in this study, such methods may serve as reference systems in scenarios where external ground truth is unavailable. UWB systems can also achieve positioning accuracy on the order of 10 cm in NLOS environments. However, any selected reference must demonstrate higher accuracy than the system being evaluated.

In summary, each experimental configuration incorporates at least one high-accuracy ground truth reference: the total station for greenhouse trials and the Vicon motion capture system for indoor trials. These references provide position data necessary for computing error metrics and validating SSSLPS performance. The following subsection details the synchronization and coordinate transformation procedures used to compare outputs across systems.

7.2.2 *Data Synchronization and Coordinate Transformation*

The SSSLPS generates position estimates within its own coordinate frame, defined by the arrangement of speakers. In typical configurations, Speaker 1 is designated as the origin (0, 0, 0) of the SSSLPS coordinate system. The positive X- and Y-axes are selected along convenient directions, such as one edge of a rectangular speaker layout. The Z-axis corresponds to the vertical direction, with Z = 0 defined by the plane of the speaker elevations. If all speakers lie at the same height, their elevation defines the Z = 0 reference plane. Otherwise, adjustments are made based on the known offset of the floor or robot operating surface.

In contrast, reference systems such as the total station and motion capture system define their own coordinate frames. The total station provides positions relative to the instrument origin, with orientation depending on how the instrument is initially aligned. The motion capture system defines its coordinate system based on camera placement and calibration. To enable direct comparison between SSSLPS output and ground truth, a coordinate transformation is performed to bring all data into a common frame. Transformation from the reference frame (e.g., total station) to the SSSLPS frame typically involves a 3D rigid-body transformation: a translation followed by a series of rotations (yaw, pitch, and roll). At least three non-collinear reference points are used to determine the transformation parameters.

Let a point P_{TS} in the total station (TS) frame be defined as (X_{TS}, Y_{TS}, Z_{TS}). Assume that Speaker 1 (P^1_{TS}), acting as the SSSLPS origin, has coordinates $(X^1_{TS}, Y^1_{TS}, Z^1_{TS})$.

The translation step aligns Speaker 1 with the origin:

$$P' = P_{TS} - P^1_{TS} = \begin{pmatrix} X_{TS} - X^1_{TS} \\ Y_{TS} - Y^1_{TS} \\ Z_{TS} - Z^1_{TS} \end{pmatrix} \tag{7.1}$$

To align the horizontal orientation, a yaw rotation is performed. The yaw angle γ is computed from the second reference point:

$$\gamma = \text{arcatan2}\left(Y_2^{'}, X_2^{'}\right) \tag{7.2}$$

Rotation about the Z-axis is applied:

$$\text{P}'' = R_z\left(-\gamma\right)\cdot\text{P}' \tag{7.3}$$

The rotation matrix $R_z(-\gamma)$ is defined as:

$$R_z\left(-\gamma\right) = \begin{pmatrix} \cos\left(-\gamma\right) & -\sin\left(-\gamma\right) & 0 \\ \sin\left(-\gamma\right) & \cos\left(-\gamma\right) & 0 \\ 0 & 0 & 1 \end{pmatrix}$$

If required, pitch adjustment is performed to account for elevation tilt. The pitch angle β is computed as:

$$\beta = arcatan2\left(Z_2^{''}, X_2^{''}\right) \tag{7.4}$$

The corresponding rotation is:

$$\text{P}''' = R_y\left(-\beta\right)\cdot\text{P}'' \tag{7.5}$$

with:

$$R_y\left(-\beta\right) = \begin{bmatrix} \cos\left(-\beta\right) & 0 & \sin\left(-\beta\right) \\ 0 & 1 & 0 \\ -\sin\left(-\beta\right) & 0 & \cos\left(-\beta\right) \end{bmatrix}$$

If necessary, roll alignment is then performed to correct tilt in the Y–Z plane. The roll angle α is calculated:

$$\alpha = \text{arcatan2}\left(Z_3^{'''}, Y_3^{'''}\right) \tag{7.6}$$

And applied via:

$$\text{P}_{\text{SSSLPS}} = R_x\left(-\alpha\right)\cdot\text{P}''' \tag{7.7}$$

with:

$$R_x\left(-\alpha\right) = \begin{bmatrix} 1 & 0 & 0 \\ 0 & \cos\left(-\alpha\right) & -\sin\left(-\alpha\right) \\ 0 & \sin\left(-\alpha\right) & \cos\left(-\alpha\right) \end{bmatrix}$$

The complete transformation is thus:

$$P_{SSSLPS} = R_x(-\alpha) R_y(-\beta) R_z(-\gamma)\left(P_{TS} - P_{TS}^{I}\right) \tag{7.8}$$

Calibration is performed by measuring the positions of at least three known reference locations (e.g., speaker positions) using the total station or motion capture system. Providing more calibration points as input can improve accuracy. Once the transformation parameters are estimated, accuracy is verified by comparing distances between additional reference points or checking the consistency of transformed coordinates.

In addition to spatial transformation, synchronization between SSSLPS and reference system timestamps is performed. The motion capture system records data at 100 Hz, and the SSSLPS PC logs timestamps using the same clock, enabling direct interpolation of ground truth for each SSSLPS update. For the total station (operating at approximately 5 Hz), timestamps are recorded via Bluetooth and aligned during post-processing. In slow-motion experiments, small timing offsets (e.g., 0.1 s) introduce negligible error. Synchronization is refined by ensuring initial positions match while the robot remains stationary. With coordinate frames aligned and time-synchronized data, point-wise error metrics, including X, Y, Z deviation as well as horizontal distance error, can be computed across matched timestamps. These metrics form the basis of performance evaluations presented in the subsequent section.

7.3 Performance Analysis and Validation

This section introduces the experimental evaluation methods used to assess the performance of the SSSLPS. Several key criteria are examined, including positioning accuracy, the influence of spatial geometry as quantified by Dilution of Precision (DOP) metrics, the effectiveness of Doppler shift compensation under motion, and the quality of signal processing based on correlation analysis. These evaluation methods are applied to both static and dynamic experiments conducted in indoor and greenhouse environments, providing a comprehensive understanding of the system's capabilities and limitations under diverse operating conditions.

7.3.1 Positioning Accuracy Evaluation

Positioning accuracy of the SSSLPS is evaluated using standard error metrics such as Mean Absolute Error (MAE) and Root Mean Square Error (RMSE), which compare estimated positions to ground truth references. Additionally, detection rate, the percentage of time the system successfully outputs a position fix, is used to measure

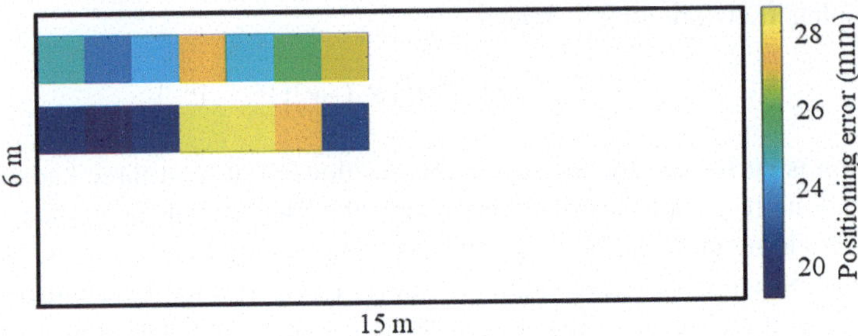

Fig. 7.7 The visualized positioning error distribution

system reliability. A high detection rate approaching 100% indicates consistent time-of-flight measurements and stable position computation.

Accuracy results are presented in both numerical and visual formats. One effective visualization involves heatmaps of positioning error across the operational area. Figure 7.7 illustrates a representative example, where color intensity corresponds to the magnitude of positioning error. Warmer colors (e.g., red, orange) indicate higher errors, while cooler colors (e.g., blue, green) represent more accurate regions. This visual format allows spatial error trends to be identified, such as elevated errors near boundaries or corners that may result from unfavorable geometry.

7.3.2 *Influence of Dilution of Precision (DOP)*

The DOP, a concept originating from GNSS, quantifies how the geometric arrangement between transmitters and receiver affects the amplification of range measurement errors. DOP serves as a multiplicative factor; the variance of positional error increases proportionally with the DOP factor due to geometric configuration. Lower DOP values correspond to favorable geometry and higher expected accuracy, whereas higher DOP values indicate suboptimal geometry and increased sensitivity to range error.

Several types of DOP are defined depending on the positional dimensions considered. Geometric DOP (GDOP) includes three-dimensional position and time. Position DOP (PDOP) addresses the three spatial coordinates. Horizontal DOP (HDOP) considers only the X and Y axes, and Vertical DOP (VDOP) pertains to the Z axis.

DOP is computed using a geometry matrix G, derived from unit line-of-sight vectors between the receiver and each speaker. For N transmitters, the matrix is defined as:

$$G = \begin{bmatrix} \frac{(x_1 - x)}{d_1} & \frac{(y_1 - y)}{d_1} & \frac{(z_1 - z)}{d_1} \\ \frac{(x_2 - x)}{d_2} & \frac{(y_2 - y)}{d_2} & \frac{(z_2 - z)}{d_2} \\ \vdots & \vdots & \vdots \\ \frac{(x_N - x)}{d_4} & \frac{(y_N - y)}{d_4} & \frac{(z_N - z)}{d_4} \end{bmatrix}$$

where (x_i, y_i, z_i) denotes the position of speaker i, (x, y, z) represents the receiver position, and d_i is the Euclidean distance between them.

The covariance matrix Q of the positioning error (under a least-squares formulation) is proportional to the inverse of the information matrix:

$$Q \propto \left(G^T G\right)^{-1}$$

From this, scalar DOP metrics are derived:

$$\text{HDOP} = \sqrt{Q_{xx} + Q_{yy}}.$$

$$\text{VDOP} = \sqrt{Q_{zz}}.$$

In the analysis for SSSLPS, HDOP is emphasized, as vertical geometry is inherently poor with speakers located on a planar ceiling. Further, axis-specific components such as x-DOP and y-DOP are introduced by extracting:

$$\text{xDOP} = \sqrt{Q_{xx}}, \text{yDOP} = \sqrt{Q_{yy}}.$$

These values indicate the directional strength of geometry along individual axes.

A DOP distribution map is generated based on the known speaker layout in the greenhouse. Figure 7.8 presents the xDOP and yDOP values across the robot's operational area. Favorable geometry (indicated by lower DOP, shown in blue) appears in some regions, whereas unfavorable configurations (high DOP, shown in yellow or red) occur at the center and periphery. The increase in DOP at the center, although counterintuitive, results from symmetry in the speaker configuration that reduces angular diversity. Higher DOP near the edges arises when the receiver becomes nearly collinear with subsets of speakers, weakening triangulation. Measured xDOP and yDOP values range from approximately 1.0 (optimal) to about 3.0 (suboptimal). These values suggest that a 1 cm range error may result in up to 3 cm of positional error in horizontal axes due solely to geometry. In contrast, zDOP values frequently exceed 10 due to minimal vertical diversity among speakers. This leads

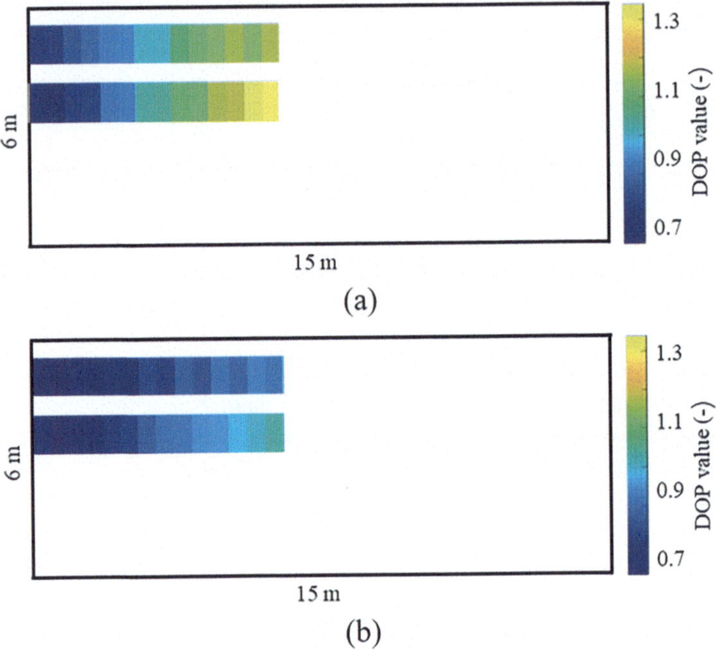

Fig. 7.8 The visualized distribution of (**a**) xDOP and (**b**) yDOP in a 15 × 6 m greenhouse

to unreliable height estimation; errors as large as 0.2–0.3 m are observed in the vertical axis.

For practical deployment, floor height constraints can be used to fix the Z value, allowing the system to focus on accurate horizontal localization. Comparison of DOP maps with the positioning error heatmap (Fig. 7.7) reveals a correlation between higher DOP regions and larger observed errors, validating the importance of geometry.

To enhance accuracy and robustness, speaker layouts can be optimized. A 3D speaker arrangement may reduce DOP across space. For example, one or more transmitters are placed at varied heights. In the greenhouse, speakers were installed near the ceiling corners; introducing a fifth speaker at a different elevation is expected to reduce zDOP and improve vertical accuracy.

7.3.3 Doppler Shift Compensation Validation

The Doppler shift effect and its compensation represent critical performance factors in the deployment of SSSLPS systems. When the receiver is in motion, the relative velocity between the transmitter and receiver induces a frequency shift in the received acoustic signal. This Doppler shift distorts the shape of the correlation peak, often resulting in degraded or failed distance estimation. Moreover, the mismatch between the transmitted and received signals reduces the effective signal-to-noise ratio of the correlation process, referred to here as *SNRcorr*.

To evaluate the impact of Doppler shift and the effectiveness of the DSC algorithm, 1D experiments are conducted under controlled conditions. These experiments are designed to isolate the effect of relative motion on distance detection and correlation integrity. Two types of setups are used to generate 1D motion: a constant-velocity test using a conveyor belt (Fig. 7.9), and an accelerating test using a sloping test bed (Fig. 7.10).

In both setups, the speaker is positioned facing the direction of motion. A microphone is mounted on the moving platform (conveyor or cart) at a fixed height of 0.2 m above the track surface. The initial distance between the speaker and microphone is set to 0.3 m, with motion continuing until the separation reaches approximately 1.9 m. In the constant-velocity configuration, the conveyor belt operates at controlled speeds This setup allows precise evaluation of system behavior at different relative velocities.

In the accelerating setup, the motion is generated by releasing a small cart down a sloped track. The slope angle is adjusted to control the magnitude of acceleration. Velocity and acceleration data are collected using a ribbon-based timer, which records positional markers on a tape affixed to the cart.

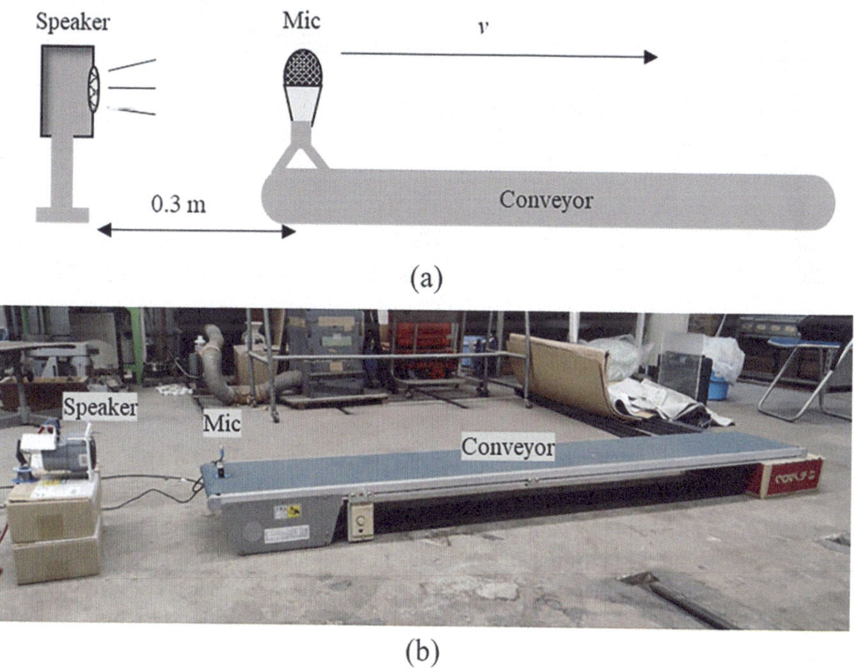

Fig. 7.9 Experimental setup for 1D measurement with constant velocity: (**a**) side view showing microphone mounted on conveyor, and (**b**) speaker orientation and test environment

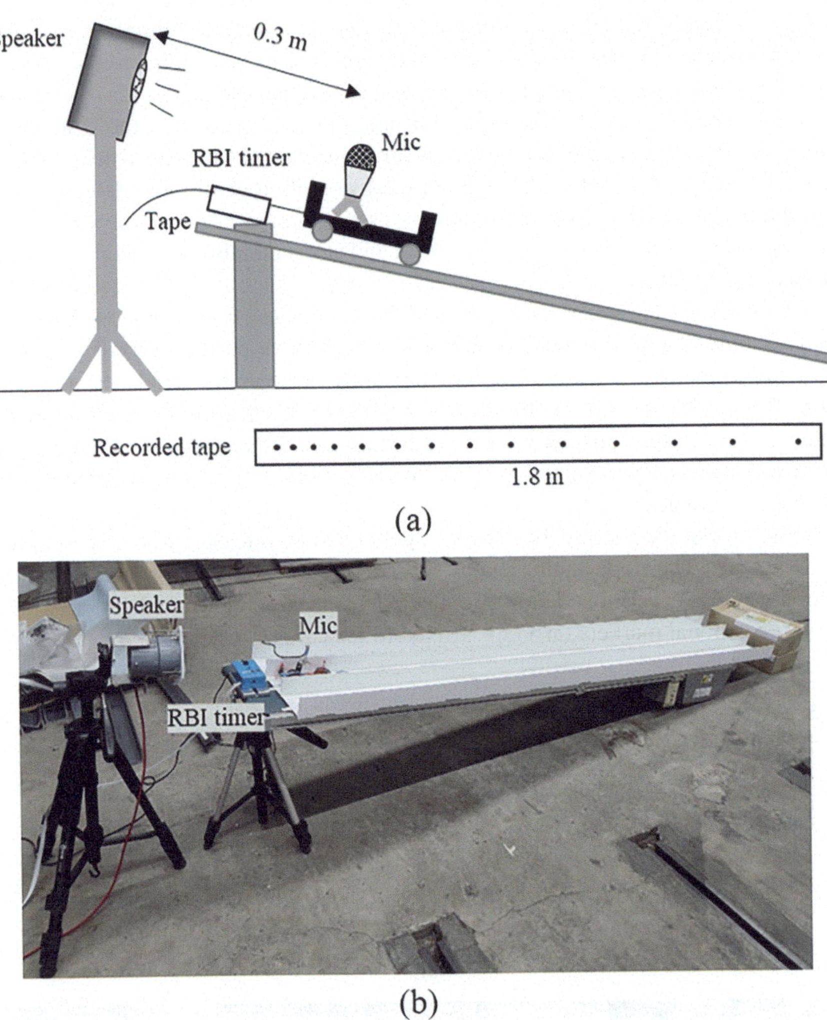

Fig. 7.10 Experimental setup for 1D measurement with acceleration: (**a**) side view of inclined test track, and (**b**) microphone mounted on rolling cart

The acceleration of the sloping testing bed was adjusted by the angle of inclination. The acceleration was provided by a recorded tape with an RBI timer.

These controlled 1D tests enable detailed analysis of the correlation performance and the effectiveness of compensation algorithms under realistic motion conditions. Results from these experiments, including correlation SNR and positioning accuracy, are used to validate the robustness of the SSSLPS under Doppler-affected scenarios.

In conclusion, this chapter presents a comprehensive validation of the SSSLPS system, demonstrating its effectiveness in achieving accurate and robust positioning performance in both greenhouse and indoor laboratory environments. Through a series of static and dynamic experiments, critical aspects such as environmental variability, multipath propagation, system geometry, and Doppler shift effects are systematically evaluated. High-precision reference systems including robotic total stations and optical motion capture are employed to provide ground truth for performance assessment. The results indicate that the SSSLPS achieves sub-decimeter accuracy and maintains a high detection rate when properly calibrated and configured. The implementation of Doppler shift compensation proves essential for preserving signal integrity during motion, while dilution of precision analysis emphasizes the importance of careful speaker placement. These findings confirm that the SSSLPS system is suitable for deployment in real-world indoor navigation and automation scenarios, and establish a basis for future enhancements through integrated sensor fusion, adaptive configuration, and extended scalability.

References

De Preter, A., Anthonis, J., Baerdemaeker, J. De. (2018). Development of a robot for harvesting strawberries IFAC-PapersOnLine *51*(17) 14–19. https://doi.org/10.1016/j.ifacol.2018.08.054

Huang, Z., Jacky, T. L. W., Zhao, X., Fukuda, H., Shiigi, T., Nakanishi, H., Suzuki, T., Ogawa, Y., & Kondo, N. (2020). Position and orientation measurement system using spread spectrum sound for greenhouse robots. *Biosystems Engineering, 198*, 50–62. https://doi.org/10.1016/j.biosystemseng.2020.07.006

Huang, Z., Ono, M., Shiigi, T., Suzuki, T., Harshana, H., Nakanishi, H., & Kondo, N. (2017). Is spread spectrum sound a robust local positioning system for a quadcopter operating in a greenhouse? *Chemical Engineering Transactions, 58*, 829–834. https://doi.org/10.3303/CET1758139

Huang, Z., Shiigi, T., Tsay, L. W. J., Nakanishi, H., Suzuki, T., Ogawa, Y., & Naoshi, K. (2021). A sound-based positioning system with centimeter accuracy for mobile robots in a greenhouse using frequency shift compensation. *Computers and Electronics in Agriculture, 187*, 106235. https://doi.org/10.1016/j.compag.2021.106235

Huang, Z., Wai Jacky, T. L., Zhao, X., Shiigi, T., Nakanishi, H., Suzuki, T., & Naoshi, K. (2019). Noise tolerance evaluation of spread Spectrum sound-based positioning system for a quadcopter in a greenhouse. *IFAC-PapersOnLine, 52*(30), 30. https://doi.org/10.1016/j.ifacol.2019.12.528

Jiang, S., Wang, S., Yi, Z., Zhang, M., Lv, X. (2022). Autonomous Navigation System of Greenhouse Mobile Robot Based on 3D Lidar and 2D Lidar SLAM. *Frontiers in Plant Science*. 13. https://doi.org/10.3389/fpls.2022.815218

Suzuki, A., Iyota, T., Yamane, A., Kubota, Y., & Watanabe, K. (2008). Accuracy on indoor position measurement by using spread Spectrum ultrasonic waves. *Transactions of the Society of Instrument and Control Engineers, 44*(6), 6. https://doi.org/10.9746/ve.sicetr1965.44.465

Tsay, L. W. J., Tomoo, S., Huang, Z., Nakanishi, H., Suzuki, T., Shiraga, K., Ogawa, Y., & Kondo, N. (2023). An acoustic based local positioning system for dynamic UAV in GPS-denied environments. *Applied Engineering in Agriculture, 39*(3), 3. https://doi.org/10.13031/aea.15397

Yan, Y., Zhang, B., Zhou, J., Zhang, Y., Liu, X. (2022). Real-time localization and mapping utilizing multi-sensor fusion and visual–imu–wheel odometry for agricultural robots in unstructured dynamic and gps-denied greenhouse environments. *Agronomy, 12*(8), 1740. https://doi.org/10.3390/agronomy12081740

Chapter 8
Applications and Future Directions

8.1 Practical Applications and Case Studies

Sound-based positioning systems have moved beyond theory into diverse real-world domains (Fig. 8.1). Thanks to their high accuracy and independence from radio-frequency spectrum, acoustic localization methods are being adopted wherever traditional positioning is unavailable or insufficient. This section surveys key application areas, ranging from controlled indoor environments such as greenhouses, factories, and warehouses to challenging conditions including underwater vehicles, and examines corresponding practical case studies. We highlight how sound-based systems meet specific needs in each domain, the technical conditions present, and the challenges encountered. These examples underscore both the current capabilities of acoustic positioning and the considerations for deploying such systems in real-world scenarios.

8.1.1 Greenhouse Robotics

Greenhouse operations have emerged as a prime application for indoor acoustic positioning. In large commercial greenhouses, autonomous robots have been used for tasks like crop monitoring, spraying, and harvesting, yet GPS is ineffective under glass roofs. A SSSLPS was developed to fill this gap, providing centimeter-level positioning accuracy for robots operating in a greenhouse. Such accuracy, combined with orientation estimates within a few degrees, enables robotic platforms to navigate planting rows precisely and perform delicate tasks in a GPS-denied environment. The SSSLPS uses ultrasonic beacons placed around the greenhouse and microphones on the robot. By measuring time-of-flight from multiple beacons, the robot can compute its 2D/3D position with high precision. These

Z. Huang, *Acoustic and Signal-Based Local Positioning*, Navigation: Science and Technology 18, https://doi.org/10.1007/978-981-95-6083-7_8

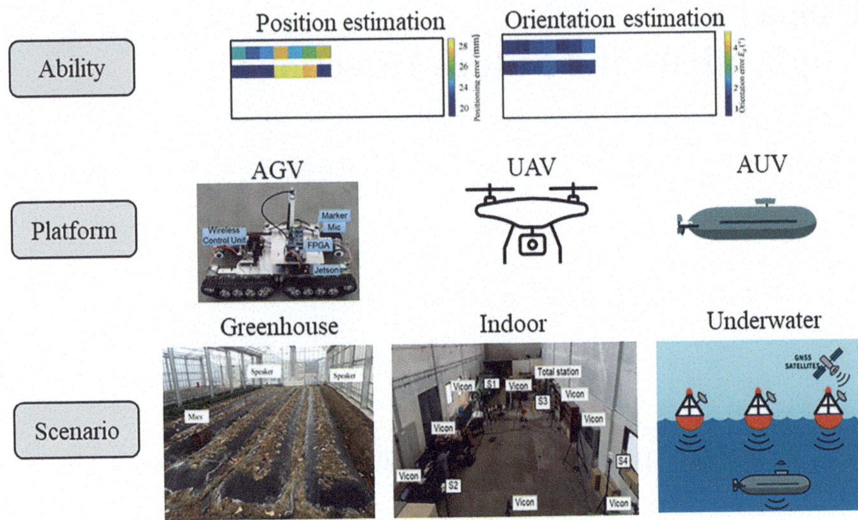

Fig. 8.1 Applications of sound-based positioning systems

results confirm that acoustic systems can meet the stringent accuracy requirements of greenhouse robotics.

The greenhouse environment presents specific technical conditions that the positioning system must handle. Temperature and humidity variations inside a greenhouse can slightly alter the speed of sound, necessitating calibration or adaptive compensation to maintain accuracy. Large fans create background noise and air currents, so the acoustic signals (often ultrasound above 20 kHz to avoid audible disturbance) must be robust against noise interference and wind. Greenhouses are generally open, rectangular spaces with line-of-sight paths along aisles, which is favorable for ultrasound propagation; however, support structures and dense foliage can introduce multipath reflections or occlusions. Researchers have addressed these challenges by using spread-spectrum chirp signals and error-correction techniques to distinguish direct arrivals from echoes. Time-division or frequency-division multiplexing is also used so that multiple robots can operate concurrently without interfering: for example, one system evaluated time-synchronized (TDMA) and frequency-separated (FDMA) acoustic bursts for tracking several robots simultaneously in a greenhouse (Tsay et al., 2022). Overall, the greenhouse case studies demonstrate that, with careful system design, sound-based positioning systems are capable of reliably guiding autonomous greenhouse robots and can achieve localization accuracy comparable to or exceeding that of other indoor positioning technologies.

8.1.2 Industrial and Warehouse Automation

Industrial indoor environments, such as factory floors and warehouses, are another domain benefitting from acoustic positioning. In these settings, autonomous guided vehicles (AGVs), forklifts, and mobile robots require sub-meter or even centimeter-level accuracy to navigate safely and efficiently among equipment and inventory. Traditional RF-based indoor localization (e.g. Wi-Fi or Bluetooth RSSI) often struggles in warehouses due to metal racks and machinery causing severe multipath and signal attenuation. Ultrasonic time-of-flight systems offer a compelling alternative: since the speed of sound is much slower than radio waves, time-of-flight measurements of ultrasound can inherently achieve higher ranging precision (on the order of millimeters) without requiring exotic hardware. Indeed, commercial solutions like Marvelmind's indoor "GPS" use ultrasound beacons to attain positioning accuracy ten times better than UWB radio in ideal conditions. This level of precision is essential in industrial applications where positioning tolerances are tight, such as aligning a robot arm to a workpiece or localizing a pallet on a high rack.

An acoustic positioning setup for a warehouse typically involves several ultrasonic transmitters mounted at fixed locations (e.g. on ceilings or walls) and receivers on the vehicles or assets to be tracked. The system measures distances via the travel time of sound pulses from beacon to receiver. Line-of-sight availability is a key factor: ultrasound cannot penetrate solid obstacles like walls or metal shelving, so the infrastructure must be planned such that each area has a direct line-of-sight to multiple beacons. Fortunately, indoor ultrasound at ~40 kHz does not penetrate walls, which can be an advantage—signals tend to stay confined to a room or area, reducing interference between zones (a property exploited in room-level localization for hospitals, discussed below). However, it also means that dense shelving or moving machinery can momentarily block the signal. One mitigation is to elevate beacons overhead to maximize visibility; another is to use a higher density of beacons so that if one path is blocked, others are available. In practice, many industrial ultrasonic systems require a mostly open line-of-sight above the factory floor or warehouse aisles.

The high refresh rate of acoustic systems (typically on the order of 1–10 Hz for each tracker) is generally sufficient for forklift or robot navigation speeds. However, scaling to a large number of moving devices requires careful coordination. Because only one ultrasonic transmission can occur within a shared space at a given moment to prevent signal overlap, systems must either employ a round-robin schedule for beacon transmissions or assign distinct ultrasonic frequencies or codes to each device. Ensuring that dozens of robots can be tracked simultaneously and in real time remains a practical challenge, which connects directly to the open research questions discussed in Sect. 8.3. Nonetheless, acoustic positioning is increasingly seen as an enabling technology for Industry 4.0 automation, complementing vision systems and RFID for a holistic indoor localization solution.

8.1.3 Smart Buildings and Infrastructure

Early smart building systems such as the Active Bat and Cricket Indoor Location System (Priyantha, 2005) (Fig. 8.2) demonstrated the potential of ultrasound for high-precision indoor localization, achieving accuracies of 1–3 cm through multilateration with ceiling-mounted transmitters and synchronized RF cues. Modern real-time location systems (RTLS), like Sonitor in healthcare, use coded ultrasonic pulses from room-mounted beacons that badges or tags detect to determine their location. Because ultrasound is confined by walls, these systems achieve reliable room-level or sub-room accuracy without complex processing, improving asset tracking, workflow efficiency, and safety. Short distances and controlled layouts allow simple time-of-flight methods to suffice, while careful synchronization prevents signal collisions.

Beyond asset tracking, acoustic localization supports context-aware environments. Microphone arrays in smart-home devices can detect a speaker's location, steer conference cameras, or adjust HVAC systems, while audible or ultrasonic cues enable low-cost positioning with existing speakers and smartphones. Though

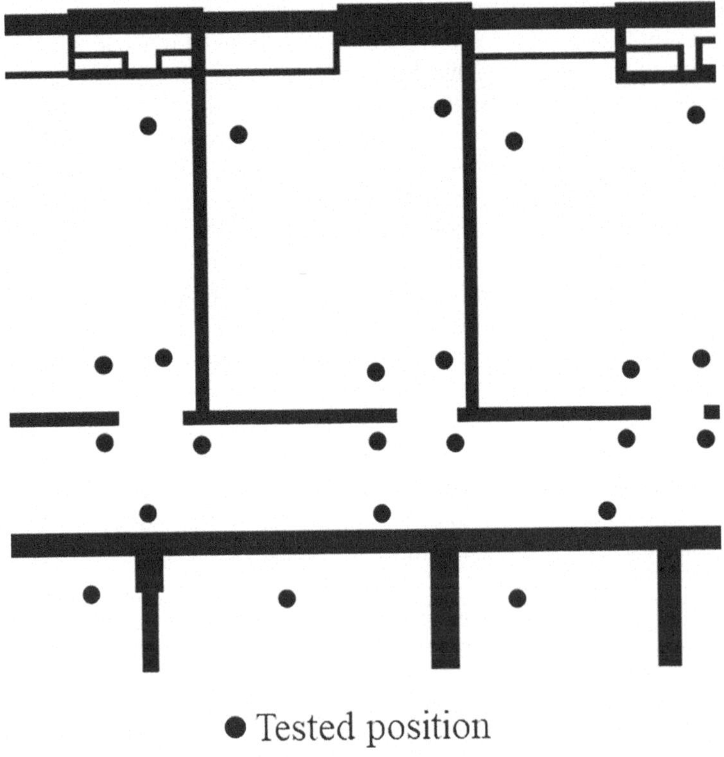

Fig. 8.2 Sound-based positioning system for indoor localization

audible approaches like the "Beep" system offer lower accuracy (~0.6 m) due to multipath and noise, they are easy to retrofit. In broader infrastructure, acoustic beacons assist navigation where GPS and radio fail, such as subway tunnels, mines, or specialized vehicle docking scenarios. By providing a self-contained, localized layer of spatial data, acoustic positioning complements vision, Wi-Fi, and 5G in the evolving sensor fusion landscape of smart buildings and cities.

8.1.4 Underwater Vehicles and Marine Robotics

Acoustic positioning is essential for underwater navigation, where GPS signals cannot propagate. Systems such as Long, Short, and Ultra-Short Baseline configurations determine the position of autonomous underwater vehicles (AUVs) by triangulating signals received at one or more fixed underwater stations. Recent designs, like Liu et al.'s distributed intelligent buoy network, combine wideband chirps and orthogonal frequency division multiplexing (OFDM) blocks to enable meter-level tracking accuracy over kilometer-scale ranges, while estimating velocity from Doppler shifts (Fig. 8.3). Another approach by Otero et al. uses GPS-equipped drifting buoys as mobile beacons, enabling rapid deployment without high-precision synchronization. Although free-floating systems risk reduced accuracy due to motion of the floating units or lack of synchronized signal reception, they provide a flexible and cost-effective alternative in environments where installing fixed infrastructure is not feasible.

Underwater acoustic positioning must contend with attenuation, multipath, Doppler, and synchronization challenges. Lower-frequency signals (10–30 kHz)

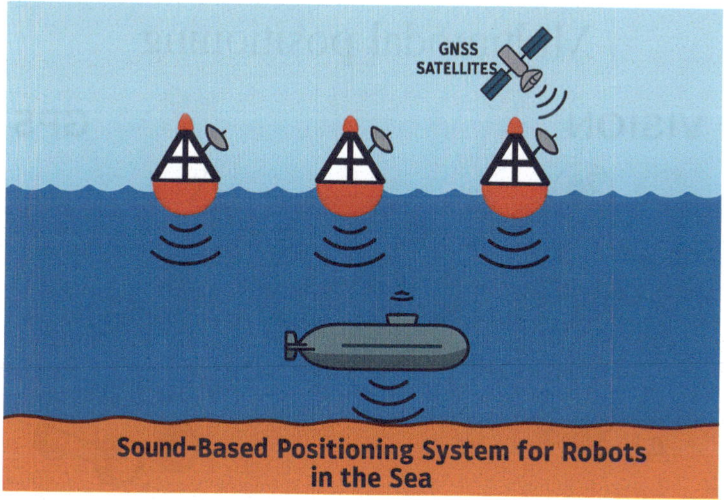

Fig. 8.3 Underwater positioning system

can travel kilometers but yield lower resolution, while high-frequency ultrasound offers centimeter-level accuracy over tens of meters. Multipath from surface and seabed reflections can distort timing measurements, mitigated by coded signals, receiver placement, or machine learning to detect direct arrivals. Doppler effects, pronounced in moving vehicles, are both a source of error and a means to estimate velocity. Despite these difficulties, acoustic positioning remains the backbone of subsea navigation in oceanography, offshore energy, and defense, with advances in algorithms and hardware steadily improving robustness under complex underwater conditions.

8.2 Integration with Emerging Technologies

As sound-based positioning systems advance in maturity, their integration with other sensing and computational technologies is becoming a central strategy for improving performance and expanding functional capabilities. Acoustic localization proves highly effective in specific domains, particularly in GPS-denied environments. However, its limitations can be mitigated by integrating complementary technologies such as computer vision, wireless navigation, and artificial intelligence (AI) (Fig. 8.4). Each modality contributes unique strengths while compensating for the weaknesses of others, creating hybrid systems that are both more robust and more accurate than any single-technology solution.

One of the most active areas of integration is with vision-based localization. Cameras, whether monocular, stereo, or depth-enabled, provide high-resolution spatial information and enable visual SLAM. Recent advances in the computer

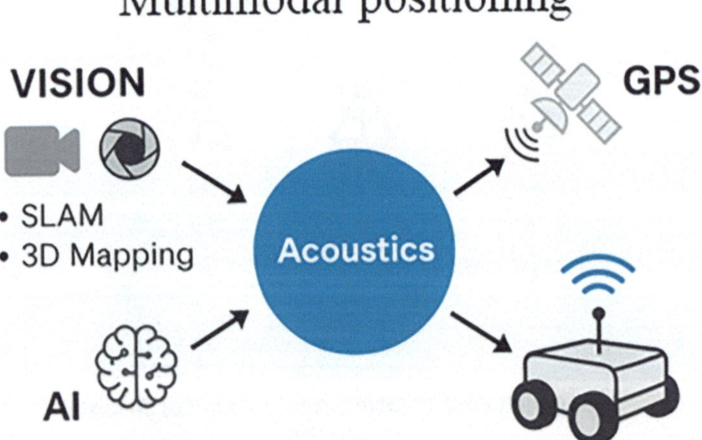

Fig. 8.4 Multimodal positioning systems

vision field have significantly improved monocular SLAM and dense 3D reconstruction capabilities, even without dedicated depth sensors. For example, Liu et al. (2025) introduced SLAM3R, a real-time monocular RGB system capable of reconstructing dense 3D point clouds at over 20 frames per second with global consistency. Such visual maps can be leveraged for positioning by matching the robot's camera observations to the prebuilt scene. However, vision-based methods are vulnerable to degraded performance in low-light conditions, low-texture environments, or when motion blur occurs. Acoustic positioning can address these weaknesses by providing drift-free absolute position updates within a fixed reference frame. In a fusion architecture, acoustic time-of-flight data anchors the visual SLAM trajectory, while the visual subsystem enriches the scene model with semantic and geometric details that acoustics alone cannot capture. An emerging research direction extends this fusion to Neural Radiance Fields (NeRF) and other learned 3D representations, where acoustic signal responses are embedded into the visual model. In such a multimodal NeRF, a robot's ultrasonic or audible chirps could be directly matched against predicted acoustic responses for that environment, potentially enabling highly precise cross-modal localization.

This synergy naturally extends to augmented reality (AR) and spatial computing applications. AR headsets and mobile AR platforms require continuous, low-latency, and high-precision tracking of position and orientation to render graphics seamlessly over the physical world. Presently, these systems rely mainly on visual-inertial odometry, which can fail in visually sparse spaces such as empty corridors or plain-walled rooms. Embedding compact acoustic transponders into the environment offers an alternative anchoring mechanism, enabling AR devices to lock onto ultrasonic beacons and stabilize tracking in challenging conditions. In multi-user scenarios, synchronized acoustic cues could ensure that all users' devices share a consistent coordinate frame. Early experimental systems already use high-frequency acoustic ranging for device-to-device localization; scaling this capability into integrated AR frameworks could substantially improve shared spatial accuracy.

Beyond vision, fusion with global navigation and wireless positioning holds promise for seamless operation across indoor and outdoor domains. A mobile robot might navigate via GPS outdoors, then transition to acoustic localization upon entering a building. The transition region, such as doorways, semi-covered areas, or partially enclosed structures, presents particular challenges because both GPS and acoustic signals may be partially degraded by multipath, NLOS propagation, or weak signal strength. Here, intelligent filtering and classification methods, often leveraging machine learning, can identify and reject unreliable data. Once low-quality measurements are excluded, advanced sensor fusion algorithms such as extended Kalman filters, particle filters, or factor graph optimizers can combine acoustic, GPS, and inertial measurements into a unified state estimate. Developing reliable indoor-outdoor interoperability remains an open problem, but ongoing work in simultaneous localization frameworks and multi-sensor calibration is closing the gap toward continuous, seamless navigation.

Perhaps the most transformative integration is the incorporation of AI and machine learning directly into the acoustic positioning process. Historically, acoustic localization relied on deterministic signal processing pipelines, including correlation for time delay estimation (Niu et al., 2023), geometric multilateration, and rule-based environmental compensation. Data-driven approaches now aim to tackle inherently complex phenomena such as multipath discrimination, self-calibration, and adaptive parameter tuning. Neural networks have been trained to identify the direct-path component of an impulse response amidst multiple echoes, learning subtle temporal, spectral, or phase-domain patterns that distinguish it from reflections. This approach has shown superior performance to fixed thresholding in reverberant or cluttered environments. At a higher level, deep learning models have been used to map raw time-of-flight inputs directly to spatial coordinates, effectively learning an environment-specific inverse localization function and bypassing explicit multilateration. The future potential of AI-enabled systems lies in their ability to adapt autonomously by continuously recalibrating for environmental changes, dynamically adjusting emission power or signal coding, and anticipating obstructions or noise events. With growing computational capacity at the network edge, even individual beacons or receivers could host lightweight AI models, enabling distributed intelligence and cooperative decision-making across the acoustic network.

In essence, integrating sound-based positioning with complementary modalities such as vision (including SLAM and NeRF), global navigation (GPS and GNSS), wireless networking, and AI transforms it from a specialized tool into a cornerstone of comprehensive localization ecosystems. Acoustics contribute reliable, geometry-based range information in GPS-denied areas; vision adds detailed mapping and semantic interpretation; wireless and GNSS provide large-scale coverage; and AI binds them together with adaptability and predictive capability. Such hybrid systems are poised to enable robust, high-precision navigation for autonomous robots, AR devices, and smart infrastructure, even in environments that currently challenge state-of-the-art localization techniques.

8.3 Vision and Future Outlook

Although sound-based positioning systems have achieved strong results, several challenges remain before they can deliver scalable, ultra-reliable performance in dynamic, real-world environments. Synchronization is a critical issue for time-of-flight ranging, as nanosecond-level timing between emitters and receivers is difficult to maintain in large or ad-hoc networks. Research is exploring self-synchronizing protocols, chip-scale atomic clocks, and one-way localization methods that solve for clock offset alongside position. Multipath interference and NLOS propagation also remain major obstacles, as reflections can mask the direct path and cause gross errors. Advanced signal processing, wideband pulses, strategic beacon placement, and machine learning classifiers offer partial solutions, but reliably resolving multipath in real time, particularly in reverberant or changing environments, remains an unsolved challenge. Environmental variability, including changes in temperature, humidity, wind, or underwater sound-speed profiles, can shift propagation speed or bend paths. Adaptive algorithms that self-calibrate from acoustic measurements, possibly aided by predictive AI models, are an emerging solution to such variations.

Scalability poses another challenge, as limited acoustic bandwidth must be shared across multiple devices. Coordinated multiple access schemes (TDMA, CDMA, FDMA) or hierarchical networks can help, but maintaining high update rates in large fleets is non-trivial, especially underwater. Cooperative localization, where robots act as beacons for each other, is promising but must avoid error propagation. Other technical hurdles include hardware-related errors (e.g., phase-center offsets, phase-wrapping ambiguities), algorithmic efficiency for embedded processors, and security, where spoofing or jamming of ultrasonic signals could disrupt navigation. Addressing these challenges will require interdisciplinary advances in signal processing, networking, AI, and secure communications, moving acoustic positioning closer to the ideal of a robust, high-precision "GPS for indoors and beyond."

Looking ahead, the future of sound-based positioning systems is bright and full of possibilities. As both the technology and its integration with other systems improve, we can envision acoustic localization becoming important in indoors as GPS is outdoors. In an analogy, just as the availability of GPS revolutionized outdoor robotics and navigation in past decades by enabling autonomous farm machinery, self-driving cars, and ubiquitous location services, a mature ecosystem of indoor acoustic positioning could drive a similar transformation for indoor robotics and smart environments.

One likely development is the increasing deployment of indoor mobile robots in roles that address labor shortages and safety concerns. For example, autonomous greenhouse robots (for seeding, pruning, harvesting) are expected to become commercially viable, particularly in regions facing agricultural labor shortfall. These robots will rely on dependable indoor positioning to operate without human oversight. Sound-based systems, possibly in combination with vision and LiDAR, will guide them along planting rows with GPS-like convenience. Similarly, we anticipate robots in hospitals, warehouses, and retail stores performing delivery and management tasks; acoustic beacons installed as part of the building's infrastructure (smart ceiling tiles emitting ultrasound, for instance) could provide a ready-made navigation grid for any compliant robot or device. Unlike relying purely on visual markers or costly laser infrastructure, acoustic positioning offers a relatively low-cost, easily maintenance solution that can be part of the building utility much like lighting or Wi-Fi.

The concept of smart infrastructure will likely extend to new levels. In the future, buildings may be constructed with built-in positioning, such as an office floor where ultrasonic emitters are embedded in the lighting system at regular intervals. From the moment a robot or a person (with a smart badge or phone) enters, they could be localized for access control, indoor navigation assistance, or environmental customization (HVAC and lighting adjusting to the occupant's location). Smart cities could deploy acoustic sensors in transportation hubs or underground facilities to guide people during emergencies when visibility is low (acoustic signals can function in smoke or darkness where vision fails). The granularity and coverage of acoustic systems are poised to improve. Current systems may pinpoint a user to within 10 cm, while future systems could narrow that to 1 cm and also provide orientation information. High-resolution acoustic phase arrays could even detect posture or gestures by analyzing the micro-Doppler of movements, enabling interfaces in which simply walking into a room in a certain way could trigger context-aware actions.

Integration with AI will push acoustic positioning into a new paradigm of intelligence and autonomy. We foresee acoustic sensors that not only locate but also characterize the environment—effectively doing double duty as echolocation devices mapping out room shapes and contents (similar to how bats use sonar to both navigate and hunt). With techniques analogous to computer vision but using sound (sometimes dubbed "acoustic vision"), a robot could potentially construct a rough 3D model of a new space by sending a few chirps and listening to the echoes. This would be immensely useful in unprepared environments or disaster scenarios. Although rudimentary forms of echo-based SLAM exist, future advancements in machine learning and signal processing could make this far more practical and detailed.

Another future direction is the standardization and interoperability of indoor positioning systems. Currently, many acoustic solutions are proprietary. As the field matures, we might see standardized acoustic beacon signals (just as Wi-Fi or Bluetooth have standards) so that any device can use a building's acoustic infrastructure. The beacons might also serve dual purposes, such as functioning as communication devices that transmit data via acoustic modems to complement Wi-Fi in IoT scenarios, or acting as environmental monitors that listen to soundscapes to detect anomalies such as glass breaks or machinery faults. In the context of multi-sensor fusion, standardized interfaces would allow seamless blending of acoustic data with inputs from future 6G wireless (which might include RF positioning), LiDAR on chips, and beyond.

We also anticipate that many current research problems will find innovative solutions. The challenge of multipath might be mitigated by future materials and design. For example, walls and floors in critical areas could be engineered with acoustic dampening or meta-material coatings that minimize reflections at the frequencies used for positioning. Ultra-wideband acoustic transducers could be developed, capable of emitting sharp impulses covering a broad spectrum, thereby improving resolution and reducing ambiguity. Miniaturization will make it feasible to include acoustic ranging capabilities in ever-smaller devices. In the future, every smartphone may not only have GPS and Wi-Fi positioning but also a tiny ultrasonic transceiver that can participate in indoor localization networks. This could lead to consumer applications that are difficult to foresee, ranging from acoustic AR gaming in a living room to personal shopping assistants that guide a user to a product shelf in a supermarket with audible pings.

Open research will evolve alongside these developments. Issues such as real-time adaptation and multi-agent coordination will likely be addressed with increased computational power and distributed algorithms, with future systems sharing processing across cloud and edge to adjust on the fly. Multi-robot systems could use game-theoretic approaches to negotiate time/frequency slots for acoustic transmissions autonomously. AI will play a role in predictive adjustment, for example by anticipating when an environment will become too noisy or when a signal path will be blocked, and by proactively switching strategies. Furthermore, as use of acoustic positioning grows, so will the need for robustness against malicious interference, potentially giving rise to a subfield of acoustic cybersecurity (ensuring the integrity of location signals).

In conclusion, the trajectory of acoustic positioning systems points toward them becoming a foundational technology in the era of pervasive robotics and smart environments. By building on current applications in greenhouses, industrial sites, underwater vehicles, and infrastructure, and by embracing integration with vision, AI, and emerging computational techniques, acoustic systems are set to dramatically extend the reach of precise positioning. In the coming years, we may witness a convergence of technologies where a robot can shift from GPS-guided outdoor navigation to ultrasonic-guided indoor navigation while also building a visual map and adapting intelligently. In such a scenario, the boundaries between indoor and outdoor environments, as well as between different sensing modalities, would no longer limit autonomous operation. Achieving this vision will require solving the open challenges discussed, but the steady progress in both research and real-world adoption is a strong indicator that sound-based positioning will play a pivotal role in the future of navigation and spatial intelligence.

References

Liu, M., Zhu, J., Pan, X., Wang, G., Liu, J., Peng, Z., & Cui, J.-H. (2023). A distributed intelligent buoy system for tracking underwater vehicles. *Journal of Marine Science and Engineering, 11*(9), 1661. https://doi.org/10.3390/jmse11091661

Liu, Y., Dong, S., Wang, S., Yin, Y., Yang, Y., Fan, Q., & Chen, B. (2025). SLAM3R: Real-time dense scene reconstruction from monocular RGB videos (No. arXiv:2412.09401). arXiv. https://doi.org/10.48550/arXiv.2412.09401.

Niu, Z., Yang, H., Zhou, L., Farag Taha, M., He, Y., & Qiu, Z. (2023). Deep learning-based ranging error mitigation method for UWB localization system in greenhouse. *Computers and Electronics in Agriculture, 205*, 107573. https://doi.org/10.1016/j.compag.2022.107573

Otero, P., Hernández-Romero, Á., Luque-Nieto, M.-Á., & Ariza, A. (2023). Underwater positioning system based on drifting buoys and acoustic modems. *Journal of Marine Science and Engineering, 11*(4), 682. https://doi.org/10.3390/jmse11040682

Priyantha, N. B. (2005). *The cricket indoor location system*. Massachusetts Institute of Technology.

Tsay, L. W. J., Shiigi, T., Zhao, X., Huang, Z., Shiraga, K., Suzuki, T., Ogawa, Y., & Kondo, N. (2022). Static and dynamic evaluations of acoustic positioning system using TDMA and FDMA for robots operating in a greenhouse. *International Journal of Agricultural and Biological Engineering, 15*(5), 5. https://doi.org/10.25165/j.ijabe.20221505.6796

The manufacturer's authorised representative in the EU is Springer Nature Customer Service Centre GmbH, Europaplatz 3, 69115 Heidelberg, Germany. If you have any concerns regarding our products, please contact ProductSafety@springernature.com

Printed and bound by CPI Group (UK) Ltd, Croydon, CR0 4YY
07/07/2026
02160926-0001